Climate Justice and Feasibility

Climate Justice and Feasibility

Normative Theorizing, Feasibility Constraints, and Climate Action

Edited by
Sarah Kenehan and Corey Katz

ROWMAN & LITTLEFIELD
Lanham • Boulder • New York • London

Published by Rowman & Littlefield
An imprint of The Rowman & Littlefield Publishing Group, Inc.
4501 Forbes Boulevard, Suite 200, Lanham, Maryland 20706
www.rowman.com

86–90 Paul Street, London EC2A 4NE

Copyright © 2021 by The Rowman & Littlefield Publishing Group, Inc.

All rights reserved. No part of this book may be reproduced in any form or by any electronic or mechanical means, including information storage and retrieval systems, without written permission from the publisher, except by a reviewer who may quote passages in a review.

British Library Cataloguing in Publication Information Available

Library of Congress Cataloging-in-Publication Data

Names: Kenehan, Sarah, 1979- editor. | Katz, Corey, editor.
Title: Climate justice and feasibility : normative theorizing, feasibility constraints, and climate action / edited by Sarah Kenehan and Corey Katz.
Description: Lanham : Rowman & Littlefield Publishers, [2021] | Includes index. | Summary: "Addresses whether and to what extent those working to better understand or achieve climate justice should think about the real-world feasibility of their theories or proposals"— Provided by publisher.
Identifiers: LCCN 2021030038 (print) | LCCN 2021030039 (ebook) | ISBN 9781538154199 (hardcover) | ISBN 9781538154212 (paperback) | ISBN 9781538154205 (epub)
Subjects: LCSH: Environmental justice. | Climatic changes—Moral and ethical aspects.
Classification: LCC GE220 .C5545 2021 (print) | LCC GE220 (ebook) | DDC 363.738/745—dc23
LC record available at https://lccn.loc.gov/2021030038
LC ebook record available at https://lccn.loc.gov/2021030039

Contents

Introduction   1
Corey Katz and Sarah Kenehan

1 Feasibility and Climate Justice   15
David A. Weisbach

2 Utopia, Feasibility, and the Need for Interpretive and Clinical Climate Ethics   33
Joshua D. McBee

3 Falling on Your Own Feasibility Sword? Challenges for Climate Policy Based on "Simple Self-Interest"   61
Stephen Gardiner and Justin Lawson

4 Climate Justice, Feasibility Constraints, and the Role of Political Philosophy   93
Brian Berkey

5 Is a Just Climate Policy Feasible?   115
Kirsten Meyer

6 The "Pathway Problem," Probabilistic Feasibility, and Non-Ideal Climate Justice   131
Jared Houston

7 Making the Great Climate Transition: Between Justice and Feasibility   155
Fabian Schuppert

8	Is Climate Justice Feasible? A Psychological Perspective on Challenges and Opportunities for Achieving a Just Climate Regime *Ezra M. Markowitz and Andrew Monroe*	173
9	Climate Change, Individual Preferences, and Procrastination *Fausto Corvino*	193
10	COVID Pandemic and Climate Change: An Essay on Soft Constraints and Global Risks *Lukas H. Meyer and Marcelo de Araujo*	213

Index 239

About the Contributors 249

Introduction
Corey Katz and Sarah Kenehan

The most recent scientific reports on climate change tell us that we are nearing a tipping point. If concentrations of greenhouse gases (GHGs) in the atmosphere are not sufficiently reduced, the world will experience dangerous global warming and all of the consequences that come with it. Despite this looming threat and decades of international negotiations, the world as a whole has had little success in reducing GHG emissions, never mind slowing their continued rise. While the 2015 Paris Agreement does set ambitious goals for minimizing warming, even if nations are able to meet their currently promised contributions toward those goals, those emissions reductions alone will not get us to the targets outlined in the agreement. But neither will those contributions represent anything close to what reasonable principles of distributive justice should require of those nations with the greatest responsibility for creating or contributing to global warming. The political and moral failures that have led to this make-or-break moment are only amplified when one considers the progress that the scholarly work on this issue has made. Since the 1980s, political philosophers and moral theorists (among others) have been exploring the moral complexities of global climate change; indeed, there is a robust and comprehensive literature detailing exactly what morality and distributive justice demand of political institutions, businesses, and individuals in the face of climate change. Yet, in the real world of climate negotiations, politics and policymaking, it is unclear how relevant or helpful these theories have been. The aim of this two-volume series is to help bridge the divide between the work of normative theorists and climate action (or inaction). In particular, contributors explore the relationship between what is "feasible" and theories and principles of climate justice, a relationship that is made even more fraught given the pressing urgency of preventing a climate catastrophe. In this, the first volume of the series, authors explore the connections (or lack thereof) between normative theorizing, normative principles, and climate action. In the second volume, authors tackle the strained relationships between climate politics, principles of justice, and concerns of feasibility.

I. BACKGROUND

Scientists have long been warning of the dangerous consequences of a rapidly warming world: intensified hurricanes, droughts and wildfires, sea-level rise, shifting ranges of diseases and illnesses, decreased biodiversity and environmental resilience, to name a few. They similarly have warned of the turmoil and suffering that was likely to result if warming continues unchecked, including, but not limited to, food and water insecurity, unplanned environmental migration, and the worsening of poverty and violence in places already replete with these difficulties. And, sadly, many of the countries most vulnerable to these harmful impacts are those who have contributed the least to the problem.

The rising global average surface temperature that is leading to these dangerous changes is the outcome of heightened concentrations of GHGs—for example, carbon dioxide, methane, nitrous oxide—in the atmosphere. These gases are produced by a wide range of human activities, from the burning of fossil fuels for energy at both industrial and consumer levels to overall land use (Boran and Katz 2017). Addressing this problem, then, requires that we find ways to reduce the concentration of GHGs in the atmosphere by slowing and eventually stopping such activities, but also that we adapt to the climate changes that are now inevitable.

The global effort to achieve both goals officially began in 1992 when the United Nations Framework Convention on Climate Change (UNFCCC) was adopted at the Rio Earth Summit. The UNFCCC came into force in 1994, which started the international process to build a multilateral climate agreement. The first Conference of the Parties (COP) under the UNFCCC convened in 1995. Negotiations at subsequent COPs led to the Kyoto Protocol, the first treaty to result from the UNFCCC process, which was adopted in 1997. The Kyoto Protocol did not require any climate action on the part of developing countries, was never signed by the United States, and most parties failed to meet the targets it set. Negotiations for a new treaty culminated in the Paris Agreement in 2015, which had a markedly different, bottom-up structure than previous efforts. The primary goal of the Paris Agreement is to limit global warming to below 2°C above pre-industrial levels, and preferably below 1.5°C (UNFCCC 2015). The agreement is based on a structure of nationally determined contributions (NDCs), whereby each nation is free to determine and submit its commitments, with periodic stocktaking and the submission of new NDCs that are expected to grow in ambition.

After the signing of the Paris Agreement, the IPCC examined the feasibility of reaching the goal of limiting warming to under 1.5°C. In the *Special Report on Global Warming of 1.5°C* (2018), researchers explained that human activities have already caused the world to warm between a range of 0.8°C and 1.2°C. At current rates of warming, we are likely to reach 1.5°C above pre-industrial levels sometime between 2030 and 2052 (IPCC 2018, 4). They further reported that the climate-related risks to health, livelihoods, food security, water supply, human security, and economic growth are likely to increase the warmer the world gets, with relative vulnerabilities to such risks increasing disproportionately for already disadvantaged populations (IPCC 2018, 9).

In spite of thirty years of global negotiations and policymaking, global GHG emissions have continued to rise and very little has been committed to adaptation funding. Moreover, the IPCC concludes that if major reductions in emissions are not achieved in ten years' time, the world as a whole will be committed to dangerous levels of warming and all of its consequences (IPCC 2018). Thus, we find ourselves in a moment of urgency with which we are frighteningly unprepared to cope. A survey of the current political landscape could lead one to think that humanity is less suited to confront the causes and effects of climate change than ever before. This is only compounded if we are not only seeking effective reductions in GHG emissions, but climate action that responds appropriately to the myriad concerns of justice in this context. To be sure, what little has been accomplished does not even approximate basic principles of climate justice, and there is a great deal of resistance to substantively addressing concerns of justice in global climate agreements.

II. CLIMATE JUSTICE AND GLOBAL CLIMATE NEGOTIATIONS

While the fundamental relation between concentrations of GHGs in the atmosphere and global warming is well understood, achieving effective and fair international coordination to respond to the problem has been far more challenging than might have been expected in the early 1990s. One of the reasons for this is that any attempt to address global climate change raises complex problems of justice. First, many of those communities that are most vulnerable to the risks and harms of climate change have contributed the least to the problem. Second, economic capacity to address the problem is not distributed equally around the globe. Third, political communities and generations have clashing interest claims in relation to the burdens of addressing climate change. These circumstances raise pressing questions about distributive justice and how to coordinate global and intergenerational cooperation (Boran and Katz 2017).

In response to such questions, normative theorists have proposed a number of principles to guide the fair distribution of the burdens associated with reducing emissions, including the polluter pays principle (Gosseries 2004; Meyer and Roser 2010; Shue 1999; Singer 2002), the ability to pay principle (Shue 1999; Caney 2005), the beneficiary pays principle (Caney 2005; Neumayer 2000; Shue 1999), and an approach that distributes equal per capita emissions among the global population (Singer 2002). Yet, it is unclear to what extent this theoretical work has influenced policymakers and policy development.

In practice, concerns about justice have sometimes been dismissed completely. Consider the "Brazil Proposal." In the late 1990s, during negotiations leading up to the adoption of the Kyoto Protocol, a group of nations put forth a proposal that identified industrialized nations as the historically responsible agents for climate change and held those agents responsible for the brunt of the costs associated with reducing emissions (UNFCCC 1997). This "Brazil Proposal" asserted a conception of fairness grounded in the polluter pays principle, a principle that is well-defended in the normative literature. Despite its solid normative grounding, the proposal was rejected by industrialized nations. This example—one of many—directs us to consider what role normative principles, theories, and theorists should and can play in policymaking.

It would be mistaken to conclude, however, that the normative work done on the issue should be dismissed as irrelevant. A concern for "equity" has always been central to both the work of normative theorists and the UNFCCC. As a basis for equitable cooperation, the UNFCCC recognizes the need to differentiate responsibility between countries in the global climate regime. The text of the UNFCCC reflects this commitment through a principle of common but differentiated responsibilities and respective capabilities (CBDR-RC) (UNFCCC 1992, Article 3). The Paris Agreement adopts this principle, but while it still calls on developed countries to take the lead in climate action, it puts in place a structure where all parties are to do their part based on their national circumstances (Boran and Katz 2017). Thus, we can understand the work of normative theorists as helping to further explain and justify existing commitments to equity and specifying their implications.

The most public example of the world of normative theorists and climate policy working together is chapter 3 of Working Group III's contribution to the IPCC's Fifth Assessment Report (AR5). The chapter, titled "Social, Economic, and Ethical Concepts and Methods," asks what *should be done* about climate change—a decidedly normative question—and uses this concern as one of the chapter's guiding inquiries (Working Group III 2014, 214). The chapter canvasses some of the most important ideas and concepts that have been developed by justice theorists over the last thirty years and includes philosophers among its lead authors. Clearly, then, work on normative theory is not completely irrelevant to climate policy or negotiations.

III. FEASIBILITY

The nature, scale, scope, and complexity of global climate change, however, test the relevance and applicability of normative theories and our ability to respond effectively to the problem. Stephen Gardiner has described the phenomenon as a "perfect moral storm." He writes, "Our problem is profoundly global, intergenerational, and theoretical. When these factors come together, they pose a 'perfect moral storm' for ethical action. This casts doubt on the adequacy of our existing institutions, and our moral and political theories" (Gardiner 2010, xii). Similarly, Dale Jamieson explains that because human moral psychology evolved when human beings lived in small, close-knit communities, "it would not be surprising if there were questions relating to anthropogenic climate change about which our everyday morality is flummoxed, silent, or incorrect" (Jamieson 2014, 147). For instance, climate change challenges our understanding of wrongful harm (Jamieson 2014, 148) and makes it difficult to ascribe fault to individuals (Jamieson 2014, 152). Both theorists likewise draw attention to the other features of the problem that make it difficult to tackle: the spatial scope of global climate change (Jamieson 2014, 161; Gardiner 2010, 24), the temporal reach of GHGs and the warming they will cause (Jamieson 2014, 165; Gardiner 2010, 32), and the fact that addressing climate change is a complex international and intergenerational collective action problem (Jamieson 2014, 162; Gardiner 2010, 24).

Jamieson and Gardiner point out the numerous feasibility constraints on achieving any action to address climate change, never mind action that comes close to reflecting the principles of climate justice that normative theorists have worked out. Given this large divide between climate policy and action and normative principles, and in recognition of the urgency of our present situation, normative theorists have started thinking about the feasibility of their proposals. To be sure, we find ourselves at a crossroads: given what is at stake, and the short time that we have to act, is it frivolous to contemplate the demands of climate justice and insist that they be represented in policy documents or action? Put differently, to what extent should those working to better understand or achieve climate justice think about the feasibility of their theories and proposals in a real-world context?

In political philosophy, reflection upon the relationship between what is feasible and normative principles and theories has sometimes taken place via a differentiation between "ideal" and "non-ideal theory." Scholars have used the term "non-ideal theory" in at least three ways (Valentini 2012). First, holding a conception of the appropriate principles of justice fixed (ideal theory), some take non-ideal theory to be about what agents who are motivated to act according to those principles should do when there are other agents who should act but aren't (Feinberg 1973; Sher 1997; Murphy 2003; Schapiro 2003; Cullity 2004, Miller 2011, Stemplowska 2016). In the case of climate change, we might wonder, for example, whether developed country parties that do seem motivated to reduce emissions (countries in the EU) should cut their emissions even more to make up for the failure of other developed countries to do so (Maltais 2014). This discussion is in some sense about feasibility, in the sense that it takes "full compliance" to be infeasible and instead develops normative principles for action in the face of the "non-compliance" of some.

Second, some take non-ideal theory to name a particular way of theorizing about justice, namely allowing relatively "sticky" institutional structures (like the international system of sovereign states) or social-scientific regularities (like the human tendency to build family groups with their children) to shape the principles of justice that are developed (Miller 2008, 2013). These elements are sometimes referred to as the "circumstances of justice," since they are taken to shape the ability of the relevant agents to accept and abide by the principles being developed (Rawls 1999a). In that sense, these elements are taken to be deep feasibility constraints not just on political action but on the very act of theorizing about justice itself. Taking seriously these concerns, Rawls worked to develop a theory of both domestic and international justice that gives us a picture of a "realistic utopia" (Rawls 1999b). The idea is that it does not make sense to, for example, theorize about what justice would be in a world without an international system of sovereign states, or spend much time thinking about what justice-improvements such a world might provide. This is because the international system of sovereign states appears to be a very sticky institutional structure.

Third, others take themselves to be non-ideal theorists insofar as they reject the need to agree upon or even have a vision of a perfectly just society in order to identify steps that could reduce existing injustice (Sen 2006, 2009; Wiens 2012). We do not need a fully formulated account of justice to serve as an end goal to pursue. Defenders of this view also claim that too much weight is put on identifying the end goal at a high level of abstraction. Unlike the previous understanding of non-ideal theory, the issue here is not with how theorists should go about developing theories or principle of justice, but with what those interested in social or political action need in hand in order to work to reduce existing injustice.

This relationship between what is feasible and "action guidance" for what we should do in the "here and now" is a topic of growing interest (Gilabert and Lawford-Smith 2012; Lawford-Smith 2013). Indeed, many contributors to this volume reflect not on normative *theorizing* about justice, but instead on the relevance of normative *principles* for guiding the urgent social and political action needed in the face of the climate crisis.

One point that many contributors highlight is that feasibility constraints can change, which can take years or may happen overnight. Thus, a morally worthwhile goal that has been dubbed infeasible may become a reality once existing constraints shift. But it is often difficult to determine the malleability of a particular constraint, or to predict the circumstances under which the constraint would be rendered less influential. Before the global COVID-19 pandemic, for instance, one might have not thought it was probable that countries would simultaneously close their borders, shut down schools, and indirectly but predictably slow their national economies by shutting down non-essential services and asking people to stay home from work. Nonetheless, in the presence of the right sort of threat, these actions not only became feasible (at least for a while) but were deemed a public health necessity. (For more on the connection between the feasibility of responding to COVID-19 and GCC, please see Meyer and de Araujo, chapter 10.) Similarly, in the context of global climate change, collective global action became more feasible with the election of American President Joe Biden in 2020 and the electoral success that allowed the Democratic Party to take control of the US Senate, which paved the way for the United States to reenter the Paris Agreement.

Even while acknowledging that feasibility constraints can and do change, it may be that in some cases, we don't have the luxury of waiting it out. In cases of extreme moral urgency, we may have to sacrifice some moral goods in order to secure an even greater moral good. Perhaps climate change is one of these instances: because of the short timeframe that we have to act in order to achieve climate stability, it could be that we have to sacrifice (to some extent, at least) what justice demands in terms of distributing the burdens of climate mitigation and adaptation.

Climate justice theorists have begun to reflect on these questions and pay more attention to the issue of feasibility over the last ten years or so (see, for instance, Posner and Weisbach 2010; Broome 2012; Caney 2014; Roser 2015; Gardiner and Weisbach 2016; Heyward and Roser 2016; Kenehan 2017). In this literature, theorists seek to strike a balance between the demands of justice and what can actually be achieved given the complexity of real-world politics and climate negotiations.

Caney (2014), for instance, provides a reflection on the issue of noncompliance in the climate justice sphere. He begins with the assumption that "it is of paramount importance that humanity avoids dangerous climate change"

(Caney 2014, 127–128). He likewise assumes noncompliance with the demands of climate justice ("first order responsibilities") and argues that we need an account that responds to this inevitability (Caney 2014, 134–135). Such an account defines "second order responsibilities," which refer "to responsibilities that some have to ensure that agents comply with their first order responsibilities" (Caney 2014, 135). As such, Caney is concerned with feasibility in a certain sense. Given the ongoing failure of many agents to meet their duties of justice in the face of climate change, Caney acknowledges the unlikeliness of agents to comply going forward and constructs a response that takes this real-world noncompliance into account.

Other climate justice theorists have not focused on the issue of noncompliance, but instead on whether normative theories and principles are needed for, or relevant to, climate action at all. Broome (2012) proposes an "efficiency without sacrifice" model for justifying climate action. On his account, by understanding GHGs as externalities, we can look to economics to remedy the problem by setting a price on GHGs and bringing them within the market (Broome 2012, 38). Importantly, Broome argues that removing these inefficiencies will not require anyone to sacrifice their interests and so does not depend on resolving conflicts of interest via principles of distributive justice. This is critical for Broome because it is clear that many nations are unwilling to make sacrifices as part of an international climate treaty. Thus, his theory is taking seriously concerns about feasibility: "Political progress over climate change has ground to almost a halt because governments are unwilling to commit their people to sacrifices. Making sacrifices unnecessary is a way to break the logjam and get the process moving again" (Broome 2012, 38).

Others have argued that traditional principles of climate justice are largely irrelevant at this point of urgency and trying to achieve climate action that meets them fails to take feasibility seriously. Rather than pursuing justice, so understood, our exclusive goal should be reducing global GHG emissions sufficiently to prevent catastrophic climate change. For example, Posner and Weisbach (2010) propose a theory called International Paretianism:

> Any treaty must satisfy what we shall call the principle of International Paretianism: all states must believe themselves better off by their lights as a result of the climate treaty. International Paretianism is not an ethical principle but a pragmatic constraint: in the state system, treaties are not possible unless they have the consent of all states, and states only enter treaties that serve their interests (Posner and Weisbach 2010, 6).

Posner and Weisbach contend that normative theorists fail to consider a serious feasibility constraint on climate action, namely that no nation will act contrary to its own interests. So, regardless of the demands of justice, a treaty that moves us toward climate stability would not be agreed upon unless it takes this constraint seriously. This might mean, for instance, that payments are made to the most prolific emitters such that they have a reason to enter a climate treaty (Posner and Weisbach 2010, 7). Gardiner (Gardiner and Weisbach 2016) has objected to this understanding, arguing, among other things, that this is climate extortion:

> Serious interference with the climate system would shape the lives of very many people around the world, and especially future generations, deeply and pervasively, to the extent of becoming at least a major determinant of their life prospects, and perhaps the dominant factor. Arguably, climate extortion exploits this fact, and this is a central part of what makes it ethically problematic (Gardiner and Weisbach 2016, 126).

As such, according to Gardiner, concerns of justice must necessarily play a role in our response to climate change. In this volume, Weisbach and Gardiner continue their debate regarding the relevance of theories and principles of climate justice, while joined by many others who take reflection upon climate justice and feasibility in multiple new directions.

IV. THEMATIC OVERVIEW

Over the course of this two-volume series, contributors respond to and expand upon the concerns we have reviewed, offer proposals that consider feasibility, and identify and explore unrecognized feasibility constraints on the pursuit of climate justice. In this volume, some authors consider the theoretical and practical relationship between feasibility constraints and normative theories and principles, while others identify novel constraints on just climate action.

To begin, contributors reflect on how we should understand the relationship between theorizing about climate justice, the principles of justice that result, and feasibility constraints. Some explore the role of theorists or the usefulness of their theories for guiding policymaking and action on climate change, while others discuss concerns with who is establishing what the feasibility constraints are and how they are doing so. If global action is urgently needed to achieve sufficient emission reductions to prevent catastrophic climate change, how should normative justice concerns come into play? Should normative theorists keep making theories? Can the principles they develop ever be met given the complexity of the climate problem? Two major voices in this debate—David Weisbach and Stephen Gardiner—continue their engagement here, joined by others who offer further nuance to these issues.

In his contribution, David Weisbach returns to his claim (sometimes made with Eric Posner) that the expectation that climate action should meet any principles of justice reduces the feasibility of achieving sufficient global emissions reductions to avoid catastrophic climate change. Such principles would require actions that many countries or people in those countries would not, under almost any circumstances, agree to. Weisbach examines this claim in light of the growing literature on the relationship between normative theory and feasibility, and reasserts that normative theories and theorizing should be left behind, at least for those who want to play a role in actually trying to address the impending climate crisis.

Joshua McBee takes up this thread and reflects on his experience as a trained normative theorist working in climate policy. He worries that deciding which policies to pursue or support based on whether they will get us closer to a highly idealized conception of justice will not only lead to policies that will gain little traction with policymakers and citizens, but that such an approach has theoretical issues as well. He concludes by making two interesting proposals for how normative theorists could effectively contribute to climate policymaking.

The next three chapters of this section claim that principles of justice and normative theorizing have a very important role to play and may even contribute to making climate action more feasible in a number of ways. Stephen Gardiner and Justin Lawson examine Weisbach's claim that "self-interest" can serve as a more feasible justification for ambitious climate action than justice claims can, and they find that argument wanting. Brian Berkey argues that there are a number of valuable practical roles that philosophical theorizing unbounded by feasibility concerns can play even in urgent circumstances like those that we currently face with respect to climate change. Kirsten Meyer agrees with Berkey and argues that the injustice of a climate proposal may make its enactment or success less feasible. She examines the "yellowjacket" movement against climate policy in France as an example.

The next two chapters are less concerned with the role that normative theories and theorizing have to play in climate action, but instead explore how we should understand feasibility assessment as a guide to action. Jared Houston suggests that, insofar as we seek guidance on climate action, we might start by setting a goal (e.g., a climate policy based on the polluter pays principle) and then doing a probability assessment of each of the different paths that might take us toward that goal. Houston argues this approach is a mistake. First, given the uncertainties inherent to both climate and economic models, accurate probability assessments become very difficult. It is hard to say which path is more feasible than another (see also McBee's chapter for a discussion of this issue). Moreover, Houston argues that thinking of feasible pathways simply in terms of climate sensitivity or economic prediction ignores the crucial aspect of political feasibility. In particular, he argues that it fails to account for coalitions in climate politics, insufficiently models the influence of opposition, and makes assumptions that bias it toward the powerful. Fabian Schuppert agrees. He argues for a crucial political aspect to feasibility assessment, and warns that, if we are not careful, feasibility judgments will reinscribe existing power structures and forms of privilege.

While the first seven chapters of this volume examine the relationship between normative theorizing, normative principles and climate action, the next two focus on identifying novel, psychological feasibility constraints on just climate action. Ezra Markowitz and Andrew Monroe focus on the way moral psychology at the level of individuals and small groups can serve as a feasibility constraint on the achievement of climate justice. They identify some of the key ways in which the nature of the climate crisis can interact with core psychological mechanisms responsible for directing morally relevant behavior. Fausto Corvino also examines psychological feasibility constraints. He argues that many who are deeply concerned about climate change continue to be insufficiently responsive to it. The reason is two psychological "short circuits" that lead to decisions and actions that do not accord with an individual's true preferences.

The volume concludes with Lukas Meyer and Marcelo de Araujo's timely reflection on feasibility, climate change, and the COVID-19 pandemic. They argue that the COVID-19 pandemic provides a sort of test-case regarding what might be feasible for the world to undertake in the face of a challenging global emergency. They analyze the feasibility constraints that stand in the way of both addressing present and future pandemics and climate change, and then argue that the constraints on addressing climate change are more difficult to overcome. Even so, they end with a note of optimism that we will all learn something about how to better address climate change in a post-pandemic world.

REFERENCES

Boran, Idil, and Corey Katz. 2017. "Climate Change Justice." *Routledge Encyclopedia of Philosophy*. Available at: https://www.rep.routledge.com/articles/thematic/climate-change-justice/v-1/sections/intergenerational-justice.

Broome, John. 2012. *Climate Matters: Ethics in a Warming World*. New York: Norton.

Caney, Simon. 2005. "Cosmopolitan Justice, Responsibility, and Global Climate Change." *Leiden Journal of International Law* 18: 747–775.

Caney, Simon. 2014. "Two Kind of Climate Justice: Avoiding Harm and Sharing Burdens." *Journal of Political Philosophy* 22(2): 125–149.

Cullity, Garrett. 2004. *The Moral Demands of Affluence*. Oxford: Oxford University Press.
Feinberg, Joel. 1973. "Duty and Obligation in the Nonideal World." *Journal of Philosophy* 70(9): 263–275.
Gardiner, Stephen. 2011. *A Perfect Moral Storm: The Ethical Tragedy of Climate Change*. Oxford: Oxford University Press.
Gardiner, Stephen, and David Weisbach. 2016. *Debating Climate Ethics*. Oxford: Oxford University Press.
Gilabert, Pablo, and Holly Lawford-Smith. 2012. "Political Feasibility: A Conceptual Exploration." *Political Studies* 60(4) (December): 809–825.
Gosseries, Axel. 2004. "Historical Emissions and Free-Riding." *Ethical Perspectives* 11(1): 36–60.
Heyward, Clare, and Dominic Roser. 2016. *Climate Justice in a Non-Ideal World*. Oxford: Oxford University Press.
IPCC (Intergovernmental Panel on Climate Change). 2018. *Global Warming of 1.5°C. An IPCC Special Report on the Impacts of Global Warming of 1.5°C Above Pre-Industrial Levels and Related Global Greenhouse Gas Emission Pathways, in the Context of Strengthening the Global Response to the Threat of Climate Change, Sustainable Development, and Efforts to Eradicate Poverty*, edited by Valérie Masson-Delmotte, Panmao Zhai, Hans-Otto Pörtner, Debra Roberts, Jim Skea, Priyadarshi R. Shukla, Anna Pirani, et al. In Press.
Jamieson, Dale. 2014. *Reason in a Dark Time: Why the Struggle Against Climate Changed Failed—And What It Means for Our Future*. Oxford: Oxford University Press.
Kenehan, Sarah. 2017. "In the Name of Political Possibility: A New Proposal for Thinking About the Role and Relevance of Historical Greenhouse Gas Emissions." In *Climate Justice and Historical Emissions*, edited by Lukas H. Meyer and Pranay Sanklecha, 198–218. Cambridge: Cambridge University Press.
Lawford-Smith, Holly. 2013. "Understanding Political Feasibility." *Journal of Political Philosophy* 21(3) (September): 243–259.
Maltais, Aaron. 2014. "Failing International Climate Politics and the Fairness of Going First." *Political Studies* 62(3): 618–633.
Meyer, Lukas, and Dominc Roser. 2010. "Climate Justice and Historical Emissions." *Critical Review of International Social and Political* 13(1): 229–253.
Miller, David. 2008. "Political Philosophy for Earthlings." In *Political Theory: Methods and Approaches*, edited Marc Stears and David Leopold, 29–48. Oxford: Oxford University Press.
Miller, David. 2011. "Taking Up the Slack? Responsibility and Justice in Situations of Partial Compliance." In *Responsibility and Distributive Justice*, edited by Zofia Stemplowska and Carl Knight, 230–245. Oxford: Oxford University Press.
Miller, David. 2013. *Justice for Earthlings: Essays in Political Philosophy*. Cambridge: Cambridge University Press.
Murphy, Liam. 2003. *Moral Demands in Nonideal Theory*. Oxford: Oxford University Press.
Neumayer, Eric. 2000. "In Defense of Historical Accountability for Greenhouse Gas Emissions." *Ecological Economics* 33(2): 185–192.

Posner, Eric A., and David Weisbach. 2010. *Climate Change Justice*. Princeton: Princeton University Press.
Rawls, John. 1999a. *A Theory of Justice: Revised Edition*. Cambridge: The Belknap Press of Harvard University Press.
Rawls, John. 1999b. *The Law of Peoples: With, The Idea of Public Reason Revisited*. Cambridge, MA: Harvard University Press.
Roser, Dominic. 2015. "Climate Justice in the Straitjacket of Feasibility." In *The Politics of Sustainability. Philosophical Perspectives*, edited by Dieter Birnbacher and May Thorseth, 71–91. London: Routledge.
Schapiro, Tamar. 2003. "Compliance, Complicity, and the Nature of Nonideal Conditions." *Journal of Philosophy* 100(7): 329–355.
Sen, Amartya. 2006. "What Do We Want from a Theory of Justice?" *Journal of Philosophy* 10(5): 215–238.
Sen, Amartya. 2009. *The Idea of Justice*. Cambridge: Harvard University Press.
Sher, George. 1997. *Approximate Justice: Studies in Non-Ideal Theory*. Lanham: Rowman & Littlefield.
Shue, Henry. 1999. "Global Environment and International Inequality." *International Affairs* 75(3): 531–545.
Singer, Peter. 2002. *One World: The Ethics of Globalization*. New Haven: Yale University Press.
Stemplowska, Zofia. 2016. "Doing More Than One's Fair Share." *Critical Review of International Social and Political Philosophy* 19(5): 591–608.
UNFCCC (United Nations Framework Convention on Climate Change). n.d. "Nationally Determined Contributsions (NDCs)." Accessed February 16, 2021. Available at: https://unfccc.int/process-and-meetings/the-paris-agreement/nationally-determined-contributions-ndcs/nationally-determined-contributions-ndcs.
UNFCCC (United Nations Framework Convention on Climate Change). 1992. "Article 3." Available at: https://unfccc.int/resource/docs/convkp/conveng.pdf.
UNFCCC (United Nations Framework Convention on Climate Change). 2015. "Adoption of the Paris Agreement." Decision 1/CP.21. Document FCCC/CP/2015/10/Add.1. Paris.
UNFCCC (United Nations Framework Convention on Climate Change), ad hoc Group on the Berlin Mandate. 1997. "Implementation of the Berlin Mandate: Additional Proposals from Parties." UNFCCC. Available at: https://unfccc.int/documents/926.
Valentini, Laura. 2012. "Ideal vs. Non-ideal Theory: A Conceptual Map." *Philosophy Compass* 7(9): 654–664.
Wiens, David. 2012. "Prescribing Institutions Without Ideal Theory." *Journal of Political Philosophy* 20(1): 45–70.
Working Group III to the Fifth Assessment Report of the Intergovernmental Panel on Climate Change. 2014. *Climate Change 2014: Mitigation of Climate Change*. Cambridge: Cambridge University Press. Available at: https://www.ipcc.ch/report/ar5/wg3/social-economic-and-ethical-concepts-and-methods/.

Chapter 1

Feasibility and Climate Justice

David A. Weisbach[1]

In a book published in 2010, Eric Posner and I argued that most theories of climate change justice suffered from two flaws (Posner and Weisbach 2010) ("PW"). First, PW argued that most theories of climate justice had internal, logical inconsistencies. They were simply badly reasoned. For example, claims based on distributive justice assumed that redistribution should occur through a climate treaty rather than through the most effective and least costly method. If redistribution through a climate treaty is less effective than other means, however, most theories of distributive justice would reject that approach. Instead, they would recommend redistributing using the more effective methods. Similar problems plagued other claims about climate justice, including claims stemming from theories of corrective justice and from theories about equality. Second, PW argued that most theories of climate justice recommended policies that were infeasible. They demanded actions that many countries or people in those countries would not, under almost any circumstances, agree to. They were utopian in the bad sense of the word.

I'm not aware of any serious grappling with the internal logical problems with the theories of climate justice that PW raise. These problems persist and are fatal to many arguments about climate change justice. Otherwise serious philosophers have been content with handy-wavy, flawed arguments or suggestions in footnotes that they have a response to PW's criticisms but simply cannot be bothered to actually say what they are (Caney 2012, n. 8). The flawed arguments remain flawed.

Since PW was published in 2010, however, there has been a substantial growth in the scholarship on the feasibility problem in political philosophy (in general, not with respect to climate change). In addition to new thinking on feasibility, much has changed in the world since 2010. Claims about feasibility depend on close attention to the facts. What is feasible depends on technologies, institutional arrangements, and the motivation and attitudes of people and countries. Almost all of these have changed since 2010.

In this chapter, I reevaluate PW's feasibility claims in light of the new literature on feasibility and the changed circumstances we now face. After reviewing PW's basic claim, I examine how changes in the world since 2010 affect feasibility concerns. I then address whether the recent arguments regarding feasibility change how we should think about the issue in the climate change context. My conclusion is that the changes in the world since 2010, for quite unfortunate reasons, only strengthen PW's concerns about the feasibility of theories of climate justice. Because of the need to reduce emissions much more rapidly now than in 2010, feasibility bites even more strongly. In addition, recent arguments about the role of feasibility in political philosophy do not have any traction in the climate justice context, at least for those who want to play a role in solving the impending climate crisis.

I. BACKGROUND

It is useful to start by recalling the reasons for a feasibility constraint on theorizing climate justice in the first place. The central reason that feasibility matters in the climate context is that most claims about climate justice are supposed to be action guiding, or at least that is how I understand them to be.[2] They are claims about what particular things people and nations should do, when, how, and in what order. They are applied philosophy. For example, Paul Baer's Greenhouse Development Rights (Baer 2013), and Peter Singer's equal allocation proposal (Singer 2002) are explicitly intended as proposals, based on a theory of justice, that can be implemented. Climate justice theorists have contributed to IPCC reports, advising governments and other actors how to respond to climate change, concluding that theories of justice are important for climate-related political decision making (Working Group III 2014, 215). Even more general normative statements, such as "past emissions should be considered in determining mitigation obligations," are, as I understand them, intended to guide action rather than be abstract statements about justice. The suggestion is that we should care about past emissions insofar as we want to determine who has the right to future emissions.

Some have responded to feasibility concerns by claiming that their discussion of climate justice aimed only to state abstract truths, condemning injustice where it was found. For example, Stephen Gardiner, when faced with feasibility problems, retreated to the claim that he was "bearing witness" to injustice (Gardiner and Weisbach 2016, 255). I take it, however, that most people who engage with issues of justice in the climate change context do so because they want to contribute to the solution to climate change. Even if some are content with bearing witness, I am not. I work on issues of climate change to help solve the problem, and for me, the question is whether theories of climate justice are useful for doing so.

To be action guiding, theories need to propose actions that people might actually take. What precisely this means is a matter of debate. It will vary by context, and it is something we can never know with certainty. If proposals lack ambition altogether, they do no work. If they deny basic facts about how people or nations behave, however, they are useless, or worse. As Joseph Carens (1996) emphasized in the immigration context, the level of fact-intensiveness or realism needs to be appropriate to the task being considered. There is no simple answer to the appropriate level of ambition.

David Estlund (2011) used a hypothetical tax proposed by Carens (1981), which I will call the "max tax" to analyze feasibility issues roughly; a max tax imposes a tax rate of 100 percent on all income and distributes the proceeds equally to all individuals, generating perfect income equality. The twist to this thought-experiment is that the max tax also includes the claim that justice or morality requires individuals to work as hard as they can notwithstanding the 100 percent tax. When people do not obey this command, they are said to be acting unjustly. That they will not obey this command is merely because they lack the appropriate motivation. But a lack of motivation alone, many argue, does not excuse failing to act (e.g., Estlund 2011). You cannot simply refuse to obey a moral command because you do not want to and escape the requirements of justice.

Contrast this with the standard approach to setting tax rates. The standard approach, developed by James Mirrlees (1971), starts off with a normative goal. In the tax context, the goal is usually something like maximizing a function of individual well-being, though other normative goals or principles of justice can be used instead.[3] Most centrally, the normative goal must have a metric that allows us to compare actions based on how close they get to the goal. The goal can be operationalized.

The theory then finds tax rates that best meet the goal assuming feasibility constraints on individual actions and feasibility constraints on the ability of the government to enforce the tax. Most centrally, the theory assumes that individuals adjust how much they work in response to taxation, calibrating the extent of the adjustments to observed behaviors. That is, it seeks to find the best—most just, based on whatever theory of justice one is using—set of tax rates, taking people as they are.

Failure to follow this procedure, that is, failure to take feasibility constraints seriously, generates tax rates that produce worse outcomes. We know this because the procedure outlined earlier finds the set of tax rates that best achieves the chosen goal given how people actually behave. A different set of rates based on how people do not actually behave will not do as well. Raising the level of ambition by raising rates and encouraging people to do more, as some feasibility theorists suggest, makes things worse, not better.

Note, moreover, that the feasibility constraints in this context are *purely* motivational. People could work harder when faced with high taxes, but they choose not to. If we want to find the best, most just policies, we cannot ignore the feasibility constraints merely because they are merely motivational rather than technological or physical limitations. That is, a system of justice that ignores feasibility constraints does worse than a system of justice that takes feasibility into account.

Another commonly discussed example of feasibility limitations is the structure of government. The design of the system of government must consider normal human motivations and actions. It must not assume that people are angels, entrusting them to do good when given power even if justice demands that they do so. James Madison and the other drafters of the US Constitution were famously concerned about this, generating the system of separation of powers in the United States government. Parliamentary democracies use other systems to constrain the actions of a powerful majority, including constitutional limitations on actions, guarantees of individual or group rights, and the ability to remove governments quickly through votes of no confidence.

Systems of government that do not consider normal human motivations can generate truly terrible outcomes. Rogues (Mao, Stalin, and so forth) that have gained power in systems without appropriate checks have led to massive suffering. Similarly, systems of government based on the idea that the motivations of citizens (as opposed to rulers) are malleable have led to devastating outcomes. And as with taxes, being a little bit more ambitious by assuming people in power or citizens will pursue justice more than they really will, may make things worse. We want to design governments for how people actually are, not how justice demands them to be.

Arguing along these lines, PW suggested that the major theories of climate justice produce infeasible recommendations and, as a result, should be discarded. Many would require transfers of resources from rich to poor nations that vastly exceed anything ever observed. Other theories would increase the costs of reducing emissions at a time when the daunting costs already make it unlikely that we will stabilize the level of CO_2 in the atmosphere at reasonably safe levels.

To illustrate, consider the theory that rich nations owe compensation to poor nations because of excessive past emissions. (For more on this idea, please see Meyer, L. in this volume.) The theory argues that rich nations used up more than their fair share of the capacity of the atmosphere to safely absorb CO_2. Rich nations, the theory argues, owe compensation for this excess use, for taking what justly belongs to someone else. As of 2016, the United States would, under a plausible interpretation of this theory, owe the rest of the world $12.4 trillion, or more than 600 years' worth of its 2016 foreign aid (Gardiner and Weisbach 2016, 225). India would have a right to receive $10.8 trillion, or more than four times its GDP in 2016. The cash flows from other rich nations would be of similar orders of magnitude. The claim that nations would make these payments is simply divorced from reality.

Some climate justice theorists get irritated when presented with these calculations, as if understanding the actual implications of a theory is a bad thing. Gardiner, for example, accuses me of trying to frighten people by presenting the numbers (Gardiner and Weisbach 2016, 253). He suggests that climate justice theorists do not mean that justice imposes such extreme obligations. But he and other theorists fail to provide a limiting principle. Analytic rigor, the hallmark of philosophy, requires one to either accept the logical implications of a claim or provide reasons why those implications do not follow. Hand-wavy denials do not suffice. If the implications of a theory are frightening, so much the worse for the theory.

In PW, we operationalized feasibility constraints by positing a theory we called International Paretianism (IP). IP starts from the premise that a feasible theory of climate justice must generate global (or almost global) participation. Because participation in a climate agreement is voluntary, nations must see it as in their interest to participate. Therefore, under IP, a feasible theory of climate justice must make each nation better off, however they perceive that.[4]

How nations perceive their own interests may vary widely, both geographically and over time. Nations will have reasonable disagreements about how different values should be weighted, including benefits and burdens in the future, and about empirical facts, for example how much some country has polluted and whether doing so was justified. Governments may aggregate the interests of their populations in different ways. They may worry about their position relative to other nations in addition to how well their own residents are doing in absolute terms. They may vary in their altruism toward people and nations who are worse off. They may have different views about how, or whether, theories of justice should affect their actions. And they may have different motivational limits: they may vary in their willingness bear certain costs.

IP attempts to incorporate all of these values, stating simply that however nations determine how to act, they must agree to participate in a climate treaty or other global climate policy. And they will only agree if they judge that the treaty or policy makes them better off by their own lights. IP was criticized as being unclear or poorly specified (e.g., Jamieson 2012; Baer 2012), and PW later attempted to refine the concept (Posner and Weisbach 2013). While I continue to believe that the concept is useful, the general feasibility concern does not depend on whether this particular formulation of the feasibility constraint is correct. Most climate justice proposals are so far outside of any plausible definition of what is feasible that the precise formulation matters little.

An important aspect of the feasibility constraint, whether it is IP or something else, is that climate justice and climate change policy more generally concerns itself with two goals. The first goal is reducing emissions and stabilizing the climate. Theories of climate justice may have implications about the level of required reductions. For example, there is a substantial literature about the discount rate (Working Group III 2014, 228–232 provides a brief survey). The discount rate determines how we should value today harms that will occur in the future and, therefore, what costs we should be willing to bear today. There is also a literature on how we should value different kinds of harms, such as harms to non-human animals or the environment (Working Group III 2014, 220). The size of these harms has implications for how much we should be willing to spend to mitigate climate change.

The second goal of theories of climate justice concerns itself with how those reductions should take place. PW called these "when, where, and how" policies (2013, 80–84). These policies include items such as which nations should reduce emissions first, how much different countries should reduce, and so forth. For example, a proposal that wealthy countries should reduce emissions first, leaving room for developing countries to increase emissions for some time, determines, in part, when, where, and how emissions reduction should take place.

A key problem for feasibility is that any choice regarding the latter—when, where, and how emissions reduction takes place—other than the lowest cost pathway raises the cost of the former—reducing emissions. The two goals conflict. In particular, because they make explicit recommendations for when, where, and how emissions reduction should take place, theories of climate justice, do not, for the most part, recommend the least cost pathway for emissions reduction. For example, it may be more expensive for wealthy countries to reduce emissions first. Allowing developing countries, or any countries, to increase emissions now may increase the cost of the necessary future reductions. This means that there is a trade-off: pursuing many theories of climate justice means making emissions reduction more expensive.

One possible way of meeting both of these conflicting goals is to implement a theory of climate justice through cash transfers, or the equivalent, and allow efficient emissions reduction after those transfers are made. Consider for example, theories based on the claim that all humans have an equal right to the remaining atmospheric sink of CO_2 (e.g., Singer 2002). These theories sometimes recommend allocating future emissions rights on a per capita basis and allowing global trading of those rights. In theory, this approach could both allocate the atmospheric sink equally while still following the least cost emissions pathway.

Many theories of climate justice do not permit such an approach. Even for those that do, however, the reality is likely different. As emphasized by Martin Weitzman (2017), bargaining over the distribution of large sums of cash or equivalents would likely interfere with bargaining over the level of emissions reduction. For example, a nation that has to transfer a large fraction of its GDP to others may be less willing to also undertake expensive emissions reductions. The theory of combining climate justice and least cost emissions reduction pathways is most likely just a theory.

As a result of these considerations, PW concluded that feasibility limitations are central to climate justice. Achieving the needed emissions reduction requires finding policies that might actually be implemented. Most theories of climate justice fail this test, and they fail it by such a wide margin that the nuances of how strict the feasibility requirement should be simply do not matter.

With this background, I turn to two tasks. The first is to consider how changes since 2010 affect the feasibility limitations on climate justice, accepting for the analysis that feasibility limitations are important. The second is to consider recent developments in the literature on feasibility limitations.

II. HOW CHANGES SINCE 2010 AFFECT FEASIBILITY

Much has changed about the climate since PW was published. The single most important fact is that, notwithstanding the Paris Agreement, emissions have continued to rise. In 2007, just before the financial crisis, global emissions of CO_2 and equivalents in other greenhouse gases were 42 gigatons/year (Mengpin and Friedrish 2020). Ten years later, in 2017, emissions had gone up more than 10 percent, to 47 gigatons/year. We have been moving in precisely the wrong direction, and are rapidly using up the remaining capacity of the atmosphere to absorb greenhouse gases (often referred to as the carbon budget).

The source of those emissions has changed as well. In 2004, China and the United States had equal emissions, of 7.1 gigatons of carbon dioxide ($GtCO_2$). By 2007, around the time PW were writing, China's emissions were 20 percent higher than US emissions: China emitted 9.1 $GtCO_2$ and the United States emitted 7.5 $GtCO_2$ in 2007. Russia and India were the distant third and fourth largest emitters, with emissions of 2.1 and 2.0 $GtCO_2$, respectively. Ten years later, in 2017, Chinese emissions had soared to 13 $GtCO_2$, while US emissions had gone down by a modest amount, to 6.5 $GtCO_2$. Meanwhile, India's emission had gone up by 50 percent to 3.0 $GtCO_2$, making it the world's third largest emitter.[5]

China and India have much larger populations than the United States does, so their per capita emissions were, and remain, much lower than US per capita emissions (Weisbach 2012). On a per capita basis, China is about average and India is below average, while countries like Australia and Canada, plus many of the OPEC nations, are at the top.[6] Greenhouse gas emissions, however, have to be stopped altogether, not just equalized on a per capita basis. This means that if one wants to understand where emissions reduction must take place, totals matter.

These changes in emissions leave dramatically less room to maneuver. According to a recent IPCC report, to stay under 1.5°C of warming emissions have to be zero, globally, in the next 20 to 30 years (IPCC 2018, 33). A similar, though slightly longer window applies for a 2°C target. Zero emissions means generating all of the world's electricity without using fossil fuels, replacing all gasoline-powered vehicles (including airplanes) with a non-emitting technology such as batteries or fuel cells, eliminating all industrial and agricultural emissions, and stopping deforestation. This will be a challenge. Any remaining emissions have to be offset by increased absorption of CO_2 into the biosphere, such as increased absorption by forests or soils, or through mechanical methods, such as capturing carbon directly from the air and storing it underground.

On the more optimistic side, wind and sometimes solar are now the cheapest form of new power generation (and in some places, they are cheaper than installed power) (US Dept. of Energy 2018, Figure 56, p. 62). They are limited because of the lack of storage technologies, so we cannot yet switch fully to these sources of energy, but the progress in lowering costs lends hope. In addition, there have been rapid improvements in electric vehicle technology. They remain a niche product but it is possible that they may soon develop into a feasible substitute for vehicles powered by internal-combustion engines. These new technologies may mean that developing nations do not need to emit as much CO_2 as previously thought in order to continue their development. Instead, they may be able to use newer, cleaner technologies to generate their needed energy. These possibilities may change what climate justice requires.

Finally, to the extent that motivation serves as a feasibility constraint, there have been shifts since 2010. Moral entrepreneurs and young people have been able to generate social movements around climate change. It is possible that we are on the cusp of a tipping point in attitudes toward climate change. Unfortunately, this is in part due to the growing evidence of its harms. In the other direction, in response to populist pressures, the United States, Brazil, Australia, and other nations have, as I write, governments that are determined to increase reliance on fossil fuels rather than reduce it. It is unclear which way the political needle will swing, but it is conceivable that we will soon have a global consensus for strong action, a consensus that we have not been able to reach before. It is also as conceivable that global negotiations will remain deadlocked or that nations will backtrack.

What do these changes mean for feasibility and climate justice more generally? The primary implication is that there is far less room to deviate from the lowest cost approach to emissions reduction. Even coming close to meeting the 2°C goal (not to speak of the 1.5°C goal) will take enormous effort and resources. Making this more expensive will make reaching these goals even less likely. The price of the justice demanded by climate justice theorists has gone up, dramatically. Feasibility concerns are more important now than ever.

III. ARGUMENTS AGAINST FEASIBILITY RESTRICTIONS

The second issue I would like to address is whether the recent, more general literature on feasibility in moral philosophy changes how we should think about feasibility in the context of climate change. Most of this literature considers feasibility restrictions in the abstract (i.e., it considers an arbitrary moral command, such as one must (or must not) engage in action x). I address here the impact of these arguments for theories of climate change justice. We can isolate four classes of arguments regarding feasibility restrictions.

The first class of argument regarding feasibility constraints is the most general, in the sense that it claims that, by definition, the principles of justice do not depend on facts. Normative claims about what is desirable are "fact insensitive." Gerald Cohen (2003) is the most prominent proponent of this view. To illustrate his argument, consider a claim that people should keep their promises because doing so helps them pursue their projects. While this claim depends on a fact—that keeping promises helps one pursue projects—we need a fact-free norm to evaluate whether pursuing projects is desirable. If that norm depends on facts, we can regress still further until we get to a fact-free norm. At bottom, evaluative statements, Cohen argues, are fact free.

Cohen explains that even someone who says that one should not do a given action because it is infeasible is ultimately committed to his fact-insensitivity claim. This is because in saying this she is tacitly accepting the claim, "one ought to take the action were it feasible," which depends on the fact-fact free norm that supports the normative "ought." Therefore, Cohen argues that even feasibility claims fall to his logic.

Cohen's argument has been subject to a number of criticisms.[7] Regardless of whether it holds generally, however, it has little applicability in the present context. Cohen's argument is about the logical structure of how one evaluates normative claims. It is not an argument about how one applies normative claims to determine what best to do. That one ought to take some action were it feasible does not tell us that we should recommend it when it is not feasible. To determine what people actually ought to do to solve a real problem, we have to consider people as they are.

In the discussion surrounding climate justice, one can see echoes of Cohen's argument that normative claims are fact-insensitive. In response to claims that his proposals were infeasible, Stephen Gardiner noted that while he hopes to guide action, he is also willing to simply "bear witness" (Gardiner and Weisbach 2016, 255). Gardiner argues that the situation he condemns is wrong regardless of whether proposed solutions to the situation are feasible. Merely bearing witness to unchecked climate change, however, would be a terrible outcome. We need to solve the problem of climate change. To do that, we need feasible proposals, not fact-insensitive normative claims.

A second, closely related class of arguments regarding the relationship of feasibility and justice is that feasibility limitations that are merely motivational are not good reasons for failing to comply with the demands of justice. For example, you cannot throw your trash in the street simply because you do not feel like disposing of it properly. Nor can you refuse to treat other people as equals simply because you prefer to unjustly discriminate. Simon Caney (2014) criticizes PW on precisely these grounds. Wealthy countries that are responsible for the majority of past emissions and that continue to have much higher per capita emissions than other countries, he argues, could comply with the demands of justice if they wanted to. They cannot simply avoid those demands because they do not want to comply.

Much of the literature on feasibility tries to map the terrain of when an action is sufficiently difficult that it should be deemed infeasible and when it is merely a motivational limitation. For example, David Estlund (2011) distinguishes between things that humans are physically incapable of doing—live without oxygen, fly, and the like—and things which they could, in some theoretical world, do with sufficient motivation, such as dance like a chicken in front of their bosses. Among other examples, he considers the max tax hypothetical discussed earlier, and Plato's theory of justice, which requires parents to surrender their infants to the community to be raised. Both of these proposals require actions that would seem to be beyond what ordinary humans can will themselves to do. Physically they are possible, but motivationally they are not.

Estlund argues that a mere motivational limitation, even ones as strong as these examples, do not block a claim that justice requires that people should do these things. Although his argument is complex, his core intuition seems to follow from the trash in the street example stated earlier: it is unacceptable to excuse people from the demands of justice simply because they do not want to comply. Pablo Gilabert and Holly Lawford-Smith (2012), David Weins (2016), and Zofia Stemplowska (2008) all offer variations on this same theme.

Some commentators, however, disagree. For example, political theorist William Galston considers Peter Singer's argument that social proximity makes no moral difference (Galston 2010, 402). Singer argues that we have equal moral obligations toward our own children, our neighbor's children, and children in distant countries. But what are we to make of the fact that no society has, or will ever, follow this prescription? The limitation is, after all, merely motivational. It is physically possible to do as Singer suggests. We could follow philosophers like Estlund and ignore the motivational limitation on Singer's proposal. Or, as Galston suggests, we could recognize that "our inability to comply with Singer's prescription suggests that something is wrong with the prescription, not with us" (Galston 2010, 402).

I cast my lot with Galston, but regardless of which position one takes on these arguments, the arguments are not relevant for proposals to implement climate change policy. Their debate is about the nature of the term "justice." It asks, what does it mean to say something is just or is required by justice, and should that statement take motivational limitations into consideration. The debate is essentially definitional in nature. It says nothing about what sort of things one might actually propose in pursuit of the demands of justice. Estlund is explicit about this: "It is important to acknowledge that any institutional proposal that ignores the facts about how people will actually tend to behave is *worthless*" (2011, 215, emphasis added). Assuming that actors will behave differently than they actually will is not a good recipe for devising proposals that one wants to be implemented.

The criminal law provides an illustration of the distinction between defining justice and pursuing a proposal for action. People commit crimes, often of terrible severity. We can, among other things, do two things. We can analyze when committing those crimes is unjust and argue that in appropriate cases, criminals have acted unjustly. Their lack of motivation to act in accordance with the principles of justice does not excuse their crimes (at least for the most part—criminal law theory explores the boundary cases). Alternatively, we can try to devise criminal laws that minimize the costs of crime, as determined by whatever principles of justice we are using. It may be appropriate to ignore motivational limitations when claiming that someone who has committed a crime has acted unjustly, but ignoring motivational limitations when devising criminal laws would be folly. Just criminal laws must be designed taking normal human motivations, including venality, into consideration.

PW took theories of climate justice to be in the latter category, to be action guiding, to be like theories of what criminal laws should look like. They are, in Estlund's terminology, institutional *proposals*. It is possible that theorists were attempting to state abstract claims about what justice entails, what Estlund calls institutional *principles*. If that is their project, feasibility concerns matter less, but it would also mean that the claims give up on any contention that they can guide action, as Estlund emphasizes (Estlund 2011).

A third more relevant line of argument regarding the relationship of justice and feasibility is the claim that we need to know the ideal outcome so that we may determine the best feasible outcome. For example, John Simmons argues that the requirement that non-ideal policies "be 'likely to be successful' requires that we know how to measure success; and that measure makes essential reference to the ultimate target, the ideal of perfect justice" (2010, 34).

Amartya Sen (2009) disagrees. He argues that we only need the ability to compare states of the world to each other, not to the ideal. Using the metaphor of a mountain, he argues that we do not need to know that Everest is the highest mountain in the world to find the highest mountain in the United States. Indeed, moving in the geographic direction of Everest—toward Nepal— might move us away from the highest domestic peak. Laura Valentini (2012, 656) similarly argues knowing the ideal does not tell us what to do in non-ideal circumstances. The choice of the best non-ideal action will be situation-specific and may be confounded by the law of the second best.

Most policy analyses proceed along lines closer to Sen's than to Simmons's. The standard way of doing policy analyses is to state an explicit goal and explicit feasibility constraints. A mathematical procedure, outlined in basic economics textbooks, known as the method of Lagrange multipliers (or in more complex settings a Hamiltonian) then allows us to find the policies that come closest to that goal within the constraints.

For example, the literature on taxation mentioned earlier uses this approach. It sets out a goal—such as maximizing a function of individual well-being—and feasibility constraints—such as the labor supply elasticity or the evasion elasticity—and then finds the set of taxes that come closest to that goal given the constraints. This approach fully accounts for second best problems because of how it incorporates feasibility constraints. It also tells us the costs of the feasibility constraints. One of the outputs of the method of Lagrange multipliers is the marginal benefit of relaxing the constraints. This is useful for understanding how much feasibility matters. Note, moreover, that we do not have to solve the optimization problem without constraints to solve it with constraints. We do not need to know the ideal state of justice. We only need a measure of "closeness" to the ideal.

As Sen emphasizes (2009, 92–102), theories of justice that do not allow comparisons of imperfect states of the world are not useful when feasibility constraints bind. Many theories of justice do not allow comparisons if the ideal cannot be reached. For example, many rights-based theories, such as Robert Nozick's libertarian set of rights (Nozick 1974), do not permit trade-offs. Similarly, theories of justice that claim that various justice-related values are incommensurable deny that we can trade one value off another.

Theories of this sort are simply unhelpful in ordinary circumstances, where feasibility constraints bind. Martha Nussbaum (2000) responds that we should understand those situations as tragic. That is true enough, often those situations are tragic, but to proceed to set plans and to come as close as we can to the demands of justice, we need to be able to make trade-offs. Ideals without trade-offs do not help.

Theories of climate justice often fail to be action guiding for this reason. They are rarely explicit about trade-offs. I am not aware, for example, of a theory that tells us how much more temperature increase we should be willing to tolerate to achieve gains in distributive justice or to reduce inequality by some amount. But without guidance for how to make trade-offs, these theories cannot be used to make climate change policy when feasibility restricts our ability to fully meet all of the stated goals (as it does).

A final line of argument on the relation of justice and feasibility is that stating ambitious and currently infeasible goals will inspire people to do more. The most widely cited examples are the various civil rights revolutions of the last fifty years. These revolutions relied on seemingly unreachable aspirations. For example, in the year 2000, few people thought that gay marriage would be a reality in the United States. The goal seemed utterly infeasible. Yet a few years later, it is legal in all fifty states, and to a great extent, accepted by the population. Similar stories hold for other civil rights. While the reasons for the rapid changes are unclear—there are often many causes for social change—aspirational claims about justice may have helped.

More generally, feasibility might be endogenous in the sense that what is feasible might be determined, in part, by what justice demands. If that is the case, we can't separately determine the feasibility constraints and apply those exogenous constraints to demands of justice. For example, Anca Gheaus (2013, 448) argues that "a willingness to apply the label 'unjust' to regrettable situations that we cannot fix is going to *advance* the action-guiding potential of a conception of justice, by providing an aspirational ideal." Behavioral economics also often makes the constraints partially endogenous: reactions to policies can depend on how they are framed. In public policy, this is sometimes referred to as shifting or expanding the Overton Window.

On the other hand, as noted, in many circumstances, trying to do more can make things worse. For example, as discussed, raising tax rates beyond the feasibility constraints produces worse outcomes, not better outcomes. The same is true for the structure of government. Shifting away from a structure that takes people as they really are, with venal motivations as well as generous, may produce terrible outcomes. The problem is more difficult when policies are multidimensional. Doing more because of broad aspirations on one dimension may mean doing less on another. Aspirational ideals might make things worse, not better.

Climate justice is precisely such a case, where climate justice theorists want both emissions reduction and other goals such as equality or good distributional outcomes. Trade-offs are inevitable. Some climate justice theorists try to avoid this problem by holding mitigation fixed and discuss how to best reach that level of mitigation (e.g., Moellendorf 2020), but this amounts to simply denying, by assumption, that there are trade-offs. Assuming away core aspects of the problem is not likely to lead to good solutions.

Indeed, trying to do too many things at once may mean none of them get done. More may mean less. For example, many theories of climate justice would require combining both a climate mitigation treaty and a massive multilateral foreign aid treaty in a single negotiation. These theories propose that we should solve the ills of colonialism and the problem of climate change all at once. As PW (2010, 86) pointed out, we do not have a multilateral foreign aid treaty, and there is no prospect for one in the near future. Trying to fashion a multilateral foreign aid treaty while simultaneously trying to solve the climate crisis may mean neither, rather than both, gets done.[8] Stating infeasible goals may be like setting infeasibly high tax rates. It may make things worse rather than better.

There is no uniform and simple answer to when ambition is good and when it is counter-productive. In the civil-rights case, ambition turned out well. In the tax case or for the structure of government, it would be counter-productive. I am not aware of a theory that helps answer the question as a general matter. As Bill Galston noted, "[t]he art of reform is to locate the outer perimeter of the desirable possible and to use it as a guide for action in the here and now" (Galston 2010, 401).

IV. CONCLUSION

Feasibility limitations remain central to understanding how theories of justice apply to climate policy. Climate policy proposals based on theories of justice that do not take feasibility limitations into consideration often will not lead to better outcomes and may make things worse (to the extent they are not simply discarded as unrealistic). Theories of climate justice can produce better results by carefully considering how feasibility limitations affect behaviors and crafting proposals that improve outcomes within those limitations.

NOTES

1. I thank Martha Nussbaum, the editors of this volume, and participants at a University of Chicago Law School workshop for comments.

2. The editor of this volume, Sarah Kenehan, indicated to me that she believes a substantial portion of the work on climate justice is ideal theory that aims to understand ideally just arrangements or principles, with these then serving as goals or standards for deciding upon what we should do. My reading of the work is different than hers. Moreover, this disagreement suggests that many in the climate justice literature have failed to be explicit about the purpose of their theorizing about climate justice, or the relation between doing so and the crucial issue of feasibility.

3. Recent work has shown how the same approach can be used to reach other goals, e.g., Saez and Stantcheva (2016).

4. PW did not include feasibility constraints that arise due to monitoring or enforcement problems. Monitoring and enforcement problems are a serious issue with many aspects of climate agreements, and any feasible climate policy needs to take these constraints into consideration as well as IP.

5. The EU is often aggregated into a single unit for purposes of counting emissions, in which case it would be the third largest emitter. The EU is made up of a large number of different countries. Many aspects of their climate polices are coordinated but others are not. It is, therefore, not clear when they should be counted as a single region and when not. I report country-level emissions for consistency.

6. In Weisbach 2012 I provided a detailed analysis of these numbers, showing that a number of small, often not-wealthy, nations are near the top of the list of per capita emitters.

7. The core criticism, well-described in Galston (2010), is that supposedly fact-insensitive principles actually do rely on facts about what it means to be human. Humans can't fly, need to breathe oxygen, have limited lifespans, reproduce sexually, and so forth. Views about human flourishing take those facts into consideration, so they are not (indeed, are never) fact-insensitive.

8. The Paris Agreement includes a small foreign-aid provision, Article 9. Article 9 requires that "Developed country Parties . . . provide financial resources to assist developing country Parties with respect to both mitigation and adaptation in continuation of their existing obligations under the Convention." This might be taken to be a counter-example to PW's claim. The size of this fund, however, is sufficiently small that it supports PW's thesis rather than countermands it. The total fund may be as high as $100 billion. While for individual humans, this seems like a lot of money, in the context of the costs of climate change, the costs of mitigation, and the demands made by climate justice theorists, the amount is just a token. For example, recall that the amount owed by the United States under backward looking theories of corrective justice was about $13 trillion, 130 times the amount in the Paris Agreement, and that is just the amount owed by one country.

REFERENCES

Baer, Paul. 2012. "Who Should Pay for Climate Change—Not Me." *Chicago Journal of International Law* 13: 507.

Baer, Paul. 2013. "The Greenhouse Development Rights Framework for Global Burden Sharing: Reflection on Principles and Prospects." *Wiley Interdisciplinary Reviews: Climate Change* 4(1): 61–71.

Caney, Simon. 2012. "Just Emissions." *Philosophy & Public Affairs* 40(4): 255–300.

Caney, Simon. 2014. "Two Kinds of Climate Justice: Avoiding Harm and Sharing Burdens." *Journal of Political Philosophy* 22(2): 125–149.

Carens, Joseph H. 1981. *Equality, Moral Incentives, and the Market: An Essay in Utopian Politico-Economic Theory*. Joseph H. Carens.

Carens, Joseph H. 1996. "Realistic and Idealistic Approaches to the Ethics of Migration." *The International Migration Review* 30(1): 156–170.

Cohen, G. A. 2003. "Facts and Principles." *Philosophy & Public Affairs* 31(3): 211–245.

Estlund, David. 2011. "Human Nature and the Limits (If Any) of Political Philosophy." *Philosophy & Public Affairs* 39(3): 207–237.

Galston, William. 2010. "Realism in Political Theory." *European Journal of Political Theory* 9(4): 385–411.

Gardiner, Stephen, and David Weisbach. 2016. *Debating Climate Ethics*. Oxford: Oxford University Press.

Ge, Mengpin, and Johannes Friedrich.2020. "4 Charts Explain Greenhouse Gas Emissions by Countries and Sectors." World Resources Institute. February 6, 2020. https://www.wri.org/blog/2020/02/greenhouse-gas-emissions-by-country-sector.

Gheaus, Anca. 2013. "The Feasibility Constraint on the Concept of Justice." *The Philosophical Quarterly* 63(252): 445–464.

Gilabert, Pablo, and Holly Lawford-Smith. 2012. "Political Feasibility: A Conceptual Exploration." *Political Studies* 60(4): 809–825.

IPCC. 2018. *Global Warming of 1.5°C: An IPCC Special Report on the Impacts of Global Warming of 1.5°C above Pre-Industrial Levels and Related Global Greenhouse Gas Emission Pathways, in the Context of Strengthening the Global Response to the Threat of Climate Change, Sustainable Development, and Efforts to Eradicate Poverty*, edited by V. Masson-Delmotte, P. Zhai, H. O. Pörtner, D. Roberts, J. Skea, P. R. https://www.ipcc.ch/sr15/.

Jamieson, Dale. 2012. "Climate Change, Consequentialism, and the Road Ahead." *Chicago Journal of International Law* 13(2): 439–468.

Mirrlees, James. 1971. "An Exploration in the Theory of Optimum Income Taxation." *Review of Economic Studies* 38: 175–208.

Moellendorf, Darrel. 2020. "Responsibility for Increasing Mitigation Ambition in Light of the Right to Sustainable Development." *Fudan Journal of the Humanities and Social Sciences*, March.

Nozick, Robert. 1974. *Anarchy, State, and Utopia*. New York: Basic Books.

Nussbaum, Martha C. 2000. "The Costs of Tragedy: Some Moral Limits of Cost-Benefit Analysis." *The Journal of Legal Studies* 29(S2): 1005–1036.

Posner, Eric, and David Weisbach. 2010. *Climate Change Justice*. Princeton, NJ: Princeton University Press.

Posner, Eric, and David Weisbach. 2013. "International Paretianism: A Defense." *Chicago Journal of International Law* 13: 347–358.

Saez, Emmanuel, and Stefanie Stantcheva. 2016. "Generalized Social Marginal Welfare Weights for Optimal Tax Theory." *American Economic Review* 106(1): 24–45.

Sen, Amartya. 2009. *The Idea of Justice*. Cambridge: Harvard University Press.

Simmons, A. John. 2010. "Ideal and Nonideal Theory." *Philosophy & Public Affairs* 38(1): 5–36.

Singer, Peter. 2002. *One World: The Ethics of Globalization*. New Haven: Yale University Press.

Stemplowska, Zofia. 2008. "What's Ideal About Ideal Theory?" *Social Theory and Practice*, August 26, 2008.

U.S. Department of Energy. 2018. "2018 Wind Technologies Market Report." https://emp.lbl.gov/sites/default/files/wtmr_final_for_posting_8-9-19.pdf.

Valentini, Laura. 2012. "Ideal vs. Non-Ideal Theory: A Conceptual Map." *Philosophy Compass* 7(9): 654–664.

Weisbach, David. 2012. "Negligence, Strict Liability, and Responsibility for Climate Change." *Iowa Law Journal* 97: 527–565.

Weitzman, Martin L. 2017. "Voting on Prices vs. Voting on Quantities in a World Climate Assembly." *Research in Economics* 71(2): 199–211.

Wiens, David. 2016. "Motivational Limitations on the Demands of Justice." *European Journal of Political Theory* 15(3): 333–352.

Working Group III to the Fifth Assessment Report of the Intergovernmental Panel on Climate Change. 2014. *Climate Change 2014: Mitigation of Climate Change*. Cambridge: Cambridge University Press.

Chapter 2

Utopia, Feasibility, and the Need for Interpretive and Clinical Climate Ethics

Joshua D. McBee

When I first started considering a career in climate policy, it seemed obvious to me that, in order to develop policy proposals, I needed to proceed as follows. First, I needed to come to a considered view about the nature of justice, abstractly conceived. Next, I needed to draw on that view to settle the character of the ideal society—which sorts of institutions it has, and roughly what sorts of policies it has in place—without worrying about whether or not there was any path from the actual world to this ideal society. Having done all that, I could set about deciding which policy proposals or institutional reforms to support by thinking about what we could do, here and now, to bring about the ideal society, or at least to get closer to that ideal. In light of the central place it gives to discussions of the ideal society, I'll call this approach to developing policy proposals "the utopian approach."[1]

My sense is that this view—found in the work of Rawls and others—is fairly widespread among philosophers.[2] I therefore suspect that many philosophers would be just as surprised as I was to learn that this is not the way most policy analysts and policymakers seem to be thinking about things. They are not thinking about utopia, and they are not reading political philosophers to get clearer about the nature of the ideal society. In fact, many, or perhaps even most, seem to think philosophy is mostly irrelevant to their work. There are exceptions to this, of course, but on the whole, my sense is that it is not common for people in the policy community either to pay much attention to philosophy or to think about policy and politics in the way the utopian approach suggests they should.

When I first realized this, my reaction was that these people are just confused about how they should be going about things—they need philosophers to set them straight! (A result of a bit too much time spent reading Plato, perhaps.) But over the next few years, as I came to better understand the needs and expectations of the policy community and the constraints within which they and especially their target audience—policymakers—have to work, I came to view my initial reaction as both arrogant and wrong-headed. In fact, I realized, these people knew something I didn't. Far from being the *best* way to develop policy proposals, I came to think that the utopian approach to policy isn't even a good one. In part this is for theoretical reasons: one version of the approach runs up against a mistake David Estlund has called "the fallacy of approximation" (Estlund 2020, 271–287). But more important for my purposes is a practical problem with the utopian approach that I imagine goes a good way toward explaining why it is not the default approach among practitioners: the justifications for particular policy proposals that this approach suggests are just too far removed from the feasibility and other practical constraints faced by policymakers to be convincing to them, and so, for understandable reasons, proposals justified in this way are unlikely to gain traction, let alone be enacted.

Though I will expand on these points later, my primary aim in this chapter is not to criticize philosophy in general or even the utopian approach in particular. I believe that, because of the prevalence of the utopian approach among philosophers, philosophy's influence within the climate policy community and, consequently, on climate policy has been more limited than it might have been.[3] Since appropriate government action is arguably indispensable to the realization of anything that might plausibly be called climate justice, I assume this is an outcome no climate ethicist wants. Even so, many philosophers seem to assume the utopian approach is the correct one. This may be because they are unaware of the difficulties with the approach that I mean to press, as I once was myself. However, I also suspect the popularity of the utopian approach stems from the sense that there is no viable alternative for philosophers who want to contribute to policy discussions.[4] With that in mind, my primary aim is to present and argue for two alternative approaches for philosophers wanting to engage in climate policy discussions. The widespread adoption of these approaches would increase the discipline's relevance to, and influence within, the climate policy community and its impact on climate policy.

I'll begin by fleshing out my criticisms of the utopian approach. Having done that, I will describe and defend two different ways philosophers might contribute to climate policy discussions. The first, interpretive climate ethics, aims to elucidate the ethical considerations relevant to decisions about which policies or institutions to support. The second, clinical climate ethics, is an approach to developing recommendations for near-term reform that eschews the utopian approach's focus on ideal worlds in favor of what we might call a "problems first" approach, one that begins, not from an idea of which ideal world we ought to realize, but from a keen sense of what has gone wrong in our own.

I. PROBLEMS WITH THE UTOPIAN APPROACH

I'll consider two forms of the utopian approach. The *ideal targets approach* treats the ideal as the end goal and focuses on identifying incremental steps we can take toward realizing the ideal; the near-term reforms it suggests are supposed to be a means to its realization. In contrast, the *ideal benchmarks approach* views the ideal as a standard against which to measure proposals for near-term reform. According to this version of the utopian approach, we ought to support those proposals for near-term reform that would bring our world as close as possible to the ideal, even if we will never realize it.[5]

As I will explain later, each of these versions of the approach has its own unique problems. In addition, however, there is a more general problem that afflicts any approach to developing and evaluating policy proposals that begins from a conception of the ideal. I begin here.

Problem 1: Difficulties about Imagining Far-Off Worlds

When we want to understand what it would be like to live in some other possible world, we have no choice but to extrapolate from the workings of our actual world. For example, if I am wondering what it would be like to live in Bozeman, I might think about my current preferences and habits to get some sense how they might translate into that new context, where different opportunities would be available to me. In cases like this, where we slightly vary one or a small number of variables, we are generally able to make more-or-less accurate predictions about what effects those changes will have, since we have often observed the effects of those or similar changes in our world. But because they are so complex, it is much more difficult to predict what will happen to entire societies as we alter them in various ways, especially as the alterations we are considering become more far-reaching. Changes in one element may prompt changes in others that we do not anticipate, and the adoption of an institution or policy meant to solve one problem may create others. Moreover, as Gerald Gaus notes, these complexities result in predictive errors that compound as we attempt to model more and more distant social worlds:

> Our models of such complex, interconnected, systems are characterized by error inflation. An error in predicting the workings of one feature will spread to errors in predicting the justice-relevant workings of interconnected features, magnifying the initial error. As this new erroneous model is used as the basis

for understanding yet further social worlds, the magnified errors become part of the new model, which is then itself subject to the same dynamic. In complex systems, small errors in predicting one variable at an early application of the model lead to drastic errors in predicting the overall system state a rather small number of iterations out (depending on the complexity ... of the system), as errors in the initial estimate of one variable both propagate to other variables and become magnified in subsequent periods (i.e., further-out social worlds). The quintessential examples of this is weather forecasting. Our predictive models of weather ten days out are drastically inferior to our models predicting tomorrow's weather. (Gaus 2016, 80)

As a result, our assessments of the character of such worlds—including, in particular, their morally relevant features—are subject to significant uncertainty. Exactly how much uncertainty will depend on how far removed from our own the worlds in question are. In extreme cases, a purportedly ideal society may differ so radically from our own that we cannot say with any confidence whether or not it would even be better than ours, let alone whether or not it would be ideal. Nor, consequently, will we be in any position to judge whether or not we ought to strive to make our world more like this purportedly ideal world (cf. Gaus 2016, 74–89). When the purportedly ideal worlds under consideration are closer to ours, we can be more confident about our judgments. However, given the manifold departures from justice that characterize our world, any plausible candidate for the ideal society is going to be quite far removed from the actual world. Some degree of uncertainty in evaluation is therefore unavoidable for all variants of the utopian approach.

The reason these dynamics matter is because they hamper the ability of proposals for near-term reform justified either by their instrumental role in bringing us closer to some purportedly ideal world (as in the ideal targets approach) or by their own relative nearness to some such world (as in the ideal benchmarks approach) to gain traction. Given the uncertainties involved in extrapolating from our world to more distant ones, proposals justified in these ways are liable to be unconvincing to policymakers, who have strong incentives to prefer proposals whose benefits are clearer.

To see the point, consider how world leaders might regard proposals intended to help us realize the future scenario that Joel Wainwright and Geoff Mann have recently suggested represents our best option for a truly just response to climate change, "Climate X" (Wainwright and Mann 2018). Wainwright and Mann decline to describe Climate X directly; instead, they explain how it differs from three alternatives, Climate Leviathan, Climate Mao, and Climate Behemoth (Wainwright and Mann 2018, chs. 2 and 8). Climate Leviathan and Climate Mao both involve the emergence of some sort of global sovereign authority, whether that be an institution, a set of institutions, or even a set of allied countries with the power to run the global table (say the United States, the European Union, and China). In contrast, both Climate Behemoth and Climate X involve a rejection of any sort of unified, global authority. Behemoth, we are told, is foreshadowed by the recent right-wing nationalist movements in Britain and the US in that led to Brexit and the election of Donald Trump, respectively (Wainwright and Mann 2018, 44–46). In addition, both Climate Behemoth and Climate Leviathan are capitalist, while Climate Mao and Climate X anti-capitalist. Climate X is thus a world in which both capitalism and the idea of a world sovereign and perhaps even international institutions like the United Nations have been rejected. In addition to this, we are given little sense what Climate X would actually look like, other than that it would be characterized by three principles: equality, the inclusion and dignity of all, and "solidarity in composing a world of many worlds" (Wainwright and Mann 2018, 175–176).

Now suppose that, in the next few years, world leaders were to have to decide—as they well might—whether or not to participate in a new, international trade agreement that helps to increase pressure on forest countries to preserve existing rainforest by coordinating trade in commodities that drive deforestation, such as cocoa, palm oil, beef, and soy. If such an agreement were to lead to reforms in the relevant industries that reduced deforestation, the effects on greenhouse gas emissions could be significant.[6] Yet, because it increases international coordination while doing nothing to dismantle capitalism, an agreement of this sort would seem to move us further from Climate X. To its proponents, therefore, this agreement will seem to prioritize near-term benefits over the presumably much greater benefits that would supposedly result from realizing Climate X. Nevertheless, it is hard to imagine the fact that it would push us further from Climate X dissuading any heads of state from signing on to a trade pact of this sort. Even in the unlikely event that they were sympathetic to Wainwright and Mann's vision, the distance between our own world and Climate X might lead heads of state to wonder, quite reasonably, whether life in Climate X would actually be better than life in other possible worlds they are now or will soon be in a position to bring about. That is, since a world without any institutions for global coordination and whose global economy is not ultimately capitalist is so different from our current world, it is hard to judge whether it would really be more just than our current world, never mind how well such a world would do at ameliorating climate change. Together with the severity of the climate crisis, these doubts are liable to make policymakers reluctant to forego the obvious near-term benefits of coordinating trade in deforestation-linked goods so as to increase their prospects of realizing Climate X.[7]

The lesson here is that, as possible worlds become further removed from our own, and the benefits of realizing them less clear, policymakers will be correspondingly less inclined to enact reforms motivated by their propensity to help us realize those worlds or judged by their relative closeness to those worlds. This is especially the case when they have open to them alternative courses of action about whose propensity to realize some tangible benefits in the near term they can be surer, as they often will. In part, this dynamic stems from the fact that policymakers must justify their decisions to constituents and—regrettably—donors. But at least for crises as dire as climate change, this focus on probable, tangible benefits would be understandable even if policymakers were motivated solely by a desire to make the world better. For when problems are this bad, policymakers will be understandably reluctant to abandon the prospect of relatively certain improvements for the sake of (maybe, one day) realizing some hazy vision of an even better world.[8]

Admittedly, the problems here will often be less severe than in the case of Climate X. After all, many of the purportedly ideal worlds explored by philosophers are both closer to our own and more fully described than Wainwright and Mann's (by way of contrast, consider, e.g., a property-owning democracy in which Rawls' two principles are satisfied). Even so, I am suggesting, these problems are to a certain extent unavoidable for utopian theorists. Regrettably, our world is far enough from any plausible candidate for the ideal society that these difficulties are likely to diminish the persuasiveness of any utopian arguments for near-term reforms to a greater or lesser degree. Even in the best case, therefore, such arguments are risky for philosophers who want to influence climate policy.

Problem 2: Uncertainties Surrounding Feasibility Assessments

With the ideal targets approach, which justifies policy proposals in terms of their tendency to help us realize the ideal, this first problem combines with a second to further weaken the proffered justifications for proposals for near-term reforms. The first problem was uncertainty about the justice of purportedly ideal worlds. The second problem is also about uncertainty, but in this case uncertainty about the *feasibility* of realizing alternative social orders, especially those that are very different from our own. In addition, this second problem tends to be more severe than the first, since it is so difficult to determine whether or not it is feasible to realize even relatively close worlds.

In order to assess the feasibility of bringing about some possible world, we must consider an *enormous* number of factors. These include both the so-called "hard constraints" which is not in our power to change, such as the laws of physics and logic, as well as the following more malleable (i.e., so-called "soft") constraints helpfully enumerated by David Wiens in the following passage:

ability constraints, which comprise facts about human abilities; *cognitive constraints*, which comprise facts about our cognitive capacity, including cognitive biases and computational limitations; *economic constraints*, which—if taken broadly—comprise facts about possible allocations of money, labour power and time; *institutional constraints*, which include facts about institutional structure and capacity (for example, the number and distribution of veto points in a collective decision procedure and the ways in which political officials are selected); and *technological constraints*, which include facts about the tools, techniques and organizational schemes available for bringing about new states of affairs. (Wiens 2015b, 453)[9]

In addition, I would add at least the following considerations:

- For elected officials, does the proposed measure conflict with any of their public commitments? Are they up for re-election soon? Might that make them more likely to support it? Less?
- Whose support is needed for the necessary reforms to happen (e.g., unions, activists, mayors and governors, key legislators, etc.)? What are the chances these individuals or groups will get on board?
- What is the state of actual or potential coalitions that might advocate for the necessary reforms? Are they well-organized? If not, what would it take to coordinate them? Do they agree on their goals? If not, how likely are they to be able to come together? Once they do come together, how durable are they likely to be?
- What is the state of public opinion on the necessary reforms? If the public does not currently support them, is there a chance opinion might shift in their favor? Either way, how much does this matter to relevant policymakers?

Finally, since many of these factors are mutable, we have to consider what the relevant timeframe is, as well as how things might change over time. If, for instance, we want to know whether it would be feasible for Congress to pass some law in the first year of a new American President's term, it will not be relevant which party might win the House in the midterms. If, however, we are concerned with what a president might be able to do before leaving office, it might.

Given how many factors must be considered, it is not surprising that it can be very difficult to determine whether or not it is feasible to bring about some alternative social order. For exactly the same reasons that modeling alternative social orders becomes more difficult as the social order in question gets further away from the actual world, assessing the feasibility of realizing some alternative social order becomes more difficult as the possible world in question becomes further away from our world. Moreover, here as there, errors at one stage compound through later stages. Consequently, as the worlds we are considering get further and further away from the actual world, our verdicts regarding the feasibility of bringing them about will be subject to greater and greater uncertainty (cf. Wiens 2015b, 467).

The reason this matters, practically speaking, is that the myriad uncertainties that plague our feasibility assessments have a tendency to undermine or weaken arguments for policy proposals that are premised on those proposals' potential to put us in a position to take the next in a series of steps on the way to realizing some purportedly ideal social order, at least in countries, like the US, where policymakers are elected and campaigns are funded by donors. Because elected officials must defend their actions to constituents and (regrettably) donors, and because many of them have relatively short terms of office, it is risky for them to throw their weight behind initiatives that may not come to fruition. What they want are proposals the benefits of which are going to be apparent in the near-term, not proposals whose benefits will materialize later on, if at all, and then only if subsequent legislative or regulatory steps are taken whose feasibility is at present unclear. To be sure, these dynamics are unfortunate for a variety of reasons (not least because of their tendency to sideline the interests of future generations). Nevertheless, they are a fact of political life. As a result, policymakers are not likely to be persuaded by a proposal that is argued for in the way the ideal targets approach suggests it should be, that is, by emphasizing its potential role in bringing about the ideal (later, maybe, if a bunch of other things go right as well).

To see how these dynamics might play out in practice, consider how heads of state and other government officials might look on Stephen Gardiner's call for a global constitutional convention focused on future generations (Gardiner 2014). Though Gardiner is clear that this is not necessarily a call to establish a world government (2014, 306–307), it is, at least, a call to establish some new international institution or institutions empowered to advocate for and safeguard the interests of future generations against the excesses of the present. Now, as I have stressed, it is very hard to know what is politically feasible right now, let alone in a few months or years—if nothing else, the pandemic-related shutdowns in March and April of 2020 taught us that much. But the prospects for creating the sort of institution or institutions Gardiner recommends do not seem good. Even at the time Gardiner published the relevant paper (2014), the likelihood that the world might come together to create any such institution or institutions was, at best, low. But today, in light of Brexit and other signs of frustration with the EU and after four years of the Trump administration withdrawing the United States from various efforts at international cooperation (the Paris Agreement, the WHO, the Iran Nuclear Deal, etc.), only the most unflappable optimist could think there is any realistic prospect of anything of the sort Gardiner proposes happening in the near future. After all, the world has not even managed to close the adaptation gap for present generations (UNEP 2018). Since each step on the path to creating the relevant institution or institutions is unlikely to be feasible, heads of state have good reason to wonder why they should take the first step Gardiner recommends, even if they agree in principle that some such institution is desirable.[10]

As with the first problem, these difficulties will be most severe when it comes to very far-off worlds like Climate X and will become correspondingly less so for worlds closer to our own. Because these problems are to a certain extent unavoidable for utopian theorists, utopian arguments for near-term reforms are inherently risky for philosophers who want to influence climate policy.[11]

How the First and Second Problem Combine to Imperil the Ideal Targets Approach

Consider what follows when we bring these two lines of thought together. Suppose we are considering some purportedly ideal social order very different from our own. Given the first problem, our attempt to determine what life in this purportedly ideal world would be like will be subject to some level of uncertainty, and so it may be unclear whether it is actually ideal, or—in extreme cases—better at all compared to the actual world. Because of the second problem, our attempts to determine whether or not it would be feasible to realize this alternative social order (or more feasible than realizing some relevant alternative) will also be subject to massive uncertainty. For both reasons, policymakers are liable to be unconvinced by arguments for near-term reform measures that appeal to their propensity to move us further down the path to realizing purportedly ideal social worlds.

The practical consequences of this point for the approach are significant. Whatever practical policy recommendations proponents of this approach can offer are supposed to be justified by their tendency to help us realize the ideal social order. The reforms they espouse are supposed to be means to that desirable end. If the foregoing is right, however, then when it comes to purportedly ideal social orders that differ radically from our own, there will be serious questions as to both whether they are in fact ideal and whether we can bring them about. Consequently, the chances that policymakers will find convincing the justifications afforded by utopian climate ethics for the policy proposals such radical visions suggest are not good. Of course, as the worlds in question become closer to our own, these problems become less severe, and the chances policymakers will be moved by the justifications available to utopian theorists become correspondingly better. As I have emphasized, however, there are limits to this dynamic, and so there is always a non-negligible risk that policy proposals grounded in the sorts of arguments the ideal targets approach suggests will not be convincing to policymakers.

Problem 3: The Fallacy of Approximation

At this point, in light of the problems with the ideal targets approach, it might be suggested that we should instead develop and evaluate proposals for near-term reform using the ideal benchmarks approach. This approach holds that we ought to adopt those among the feasible, near-term reforms that most closely approximate the ideal. However, this version of the utopian approach is no less problematic than the ideal targets approach. As I will explain in a moment, the problems are in part very similar to those with the ideal targets approach. In this case, however, there is also a more theoretical problem: the ideal benchmarks approach rests on what David Estlund has recently called the "fallacy of approximation" (Estlund 2020, 271–287).

Consider three sets: first, some set of valuable conditions, S; second, a subset of S, ss; and third, a subset of S that is a proper superset of ss, ss' (i.e., ss' consists of all of the conditions in ss and, in addition, at least one further condition from S). The fallacy, Estlund says, consists in reasoning as though the satisfaction of ss' were, of necessity, more valuable than that of ss (Estlund 2020, 274). The issue with this way of reasoning is simple: the value of some set of conditions depends on its composition, not just how close it is to some valuable set (cf. Estlund 2020, 275). For example, suppose it would be good for you to have a bike (so that you could use it for exercise, say). It does not follow that it would be good to have a tire, since the tire by itself will not enable you to ride a bike for exercise. Nor, for the same reason, does it follow that it would be better to have a tire and a chain than to have the tire alone: these components derive what value they have from the contribution they make to a valuable whole and are useless on their own. Nor, finally, does it follow that, if you want a bike but only have the option to buy a tire and chain, you ought to buy both rather than one or neither. Unless you are in a position to acquire all the other components and have the time and ability to assemble them correctly, these components will do you no good.

This is bad news for the ideal benchmarks approach, which recommends adoption of those proposals that would allow us to realize the nearest feasible approximation of the ideal. The thought behind this approach is that the fact that some social order, O_1, more closely approximates the ideal than some other, O_2, entails that that O_1 is better, or more just, than O_2. But as we can now see, this is an instance of the fallacy of approximation: in fact, O_1 may not be at all better or more just than O_2. The method recommended by the ideal benchmarks approach is therefore not at all a reliable way of identifying near-term reforms that will make the world more just.

Suppose, for example, that I am thinking about what can be done at the policy level to facilitate easier movement across national boundaries for people displaced by climate change, and suppose I am persuaded by Arash Abizadeh's provocative argument in his "Democratic Theory and Border Coercion." Abizadeh argues that the current border regime is inconsistent with a democratic view of political legitimacy that sees the exercise of political power as legitimate if, and only if, "it is actually justified by and to the very people over whom it is exercised, in a manner consistent with viewing them as free (autonomous) and equal" (Abizadeh 2008, 41). He suggests that, instead of the current regime in which states take themselves to rightly exercise unilateral control of cross-border movement and citizenship, this democratic view of political legitimacy entails that justification for a regime of border control is owed both to members and nonmembers of the controlling state. Abizadeh's argument for this striking conclusion is, as he puts it, "surprisingly simple":

> First, a democratic theory of popular sovereignty requires that the coercive exercise of political power be democratically justified to all those over whom it is exercised, that is, justification is owed to all those subject to state coercion. Second, the regime of border control of a bounded political community subjects both members and nonmembers to the state's coercive exercise of power. Therefore, the justification for a particular regime of border control is owed not just to those whom the boundary marks as members, but to nonmembers as well. (ibid., 45)[12]

Now, it seems clear that this argument suggests immigration and naturalization policies ought to be decided, not by individual states, but by a global governing body comprised of representatives from every country in the world. Obviously, however, it is not currently feasible for the world to transition directly to this radically different border regime. For what country currently stands ready to cede territorial sovereignty to better respect the autonomy of foreigners? And what current citizenry is so broad-minded that they would support a politician who advocated this? Given that this is so, the ideal benchmarks approach suggests we ought to support adoption of the nearest feasible approximation of the sort of border regime Abizadeh's argument suggests. It is, of course, hard to say what that might look like, but my guess is that the closest we are likely to be able to get to the border regime Abizadeh's argument suggest is a set of bilateral or multilateral agreements for countries to set immigration and naturalization policy together. In principle, these sorts of arrangements could be perfectly egalitarian, with each party to these agreements having an equal say in the relevant policies. In practice, however, a scenario in which more powerful nations are generally able to set the terms in such a way as to preserve or increase their relative advantages seems more likely. Suppose that is right. Would such a border regime be better than the present one, in which each nation sets its own immigration and naturalization policies and generally gives preferential treatment to those immigrants whose presence would benefit it in some way? Not obviously; in fact, it seems unlikely to differ much at all from the present border regime.

At this point someone might object that, while this is true enough, the fact that it is not a reliable way of identifying reforms that will make things better is not really a very compelling critique of the ideal benchmarks approach. Consider again the case of the would-be bicycle-builder. It is true that having only a tire and a chain is not better than having no parts at all if you are not in a position to acquire all the other components and if you don't have the time and ability to assemble them correctly. But, typically, it will be possible to acquire the additional parts, acquire any necessary skills, and make time to put everything together. In *that* circumstance, it *is* better to have two parts than to have one or none, since that means you have less to do before you achieve your goal.

Similarly, it might be suggested, it is better to try to build the sort of egalitarian migration governance regime Abizadeh envisions even if we end up in the aforementioned situation, since we would then be closer to realizing (what I'm supposing for the sake of argument is) an ideally just migration governance regime. Even if the initial results are no better than our current arrangements, it might be suggested, we could always try later to improve things, for instance by adding more countries, or by finding ways to even the playing field between countries with vastly different amounts of power in negotiations. And just as it is better to have two bike parts rather than none, having made some initial efforts toward the intended governance regime would be better than not having done so. Getting part of the way toward the goal reduces the amount of work remaining to be done. Consequently, it might be suggested, even though the ideal benchmarks approach rests on the fallacy of approximation, we still have reason to argue for policy proposals in this way. Even if the reforms will not immediately make us a better off, they will reliably put us in a better position to do so later on, and that is a justice-improvement.[13]

The objection is fair. However, I doubt that the fact that we may be able to improve things later is really of much help to the ideal benchmarks approach. Because it rests on the fallacy of approximation, the ideal benchmarks approach is liable to suggest adoption of near-term reforms that will not in fact make us better off, meaning that the odds are good that policymakers will have some reservations about embracing the measures it recommends. One could imagine these concerns being assuaged by the prospect that adopting these reforms might put us in a better position to adopt further reforms that would make us better off, as the objection suggests. However, the uncertainty that surrounds any assessment of the feasibility of subsequent reforms means that policymakers are unlikely to find this a compelling response to their concerns, especially given that they might also harbor doubts about whether the purportedly ideal world the reforms are supposed to get us closer to would actually be ideal. In the end, then, the trouble with the ideal benchmarks approach is much the same as with the ideal targets approach: policymakers are just not likely to be persuaded by the kinds of arguments the ideal benchmarks approach suggests, at least in contexts where they are answerable to voters and donors.

The Upshot

Let us pause and take stock. A general problem for the utopian approach, I said, was that our assessments of the character of purportedly ideal social orders are characterized by greater and greater uncertainty as those social orders become further and further removed from our own. Climate X provides a particularly extreme illustration of this problem, but it is relevant for any attempts to motivate near-term reforms by pointing to purportedly ideal worlds, given how far removed from the actual world any plausible candidate for the ideal would have to be. In addition, both versions of the approach run up against a similar problem having to do with feasibility assessments. Given how many factors such assessments must consider and how liable they are to change over time, it will typically be unclear whether or not the subsequent steps on the path toward realizing the ideal beyond the first will be feasible. As a result, the arguments for particular proposals the utopian approach suggests are liable to be unconvincing to policymakers, regardless of which of the two versions of the approach those arguments draw on.

Taken together, these considerations show that neither version of the utopian approach is a good way for philosophers to develop and argue for policy proposals, at least if their goal is to influence climate policy by persuading policymakers to take up certain proposals over others. They also suggest a more general lesson for philosophers seeking to influence policy. At root, the problems with each version of the utopian approach stem from the fact that they seek to justify policy proposals by pointing beyond themselves to some purportedly ideal world. What policymakers want to see, and what philosophers wanting to influence them ought to try to provide, are proposals the enactment of which is desirable in its own right, proposals they can justify supporting and whose enactment they can claim as a victory. What they do not want, and will not find helpful, are proposals for reforms that *might* contribute to the realization of a world that *might* be ideal, or that would make our world more closely approximate that possibly ideal world, even if that would not make our world better.

Even so, I do not think utopian thinking is pointless. While the utopian approach is a bad way to develop and argue for policy proposals, utopian thinking can still serve several important functions. First, by helping to make clear just how far apart the worlds our values suggest to us are from the world we live in, utopian thought allows us to "bear witness" to the inadequacy of the world as currently constituted, thereby making clear, to both present and future generations, that we recognize our shortcomings (cf. Morrow 2020, 77; Gardiner and Weisbach 2016, 255). Second, utopian thinking can help open people's minds to different ways of doing things or help to persuade them that alternatives are more reasonable than they have hitherto thought. To the extent that lack of awareness of, inability to imagine, or dismissiveness toward different ways of organizing society are themselves barriers to their realization, utopian thinking can itself help to break down the feasibility constraints that stand in the way of efforts to bring about better worlds. In addition, and relatedly, utopian thinking can help us to see the present social order differently, and in some cases more clearly. By thinking about how the world ought to be, we may not only get clearer about why various features of our world are problematic but also come to see as problematic features of our world that had not previously seemed to be. Finally, because it helps us to more clearly envision a better world, utopian thought can both help to clarify the barriers that currently make realizing some purportedly ideal social world infeasible and inspire attempts to break them down.

II. TWO ALTERNATIVE APPROACHES TO CLIMATE ETHICS

By now, many readers will have wondered what the alternative is for philosophers who want to influence policymakers, policy analysts, and advocates. My aim in the remainder of this chapter is to describe and defend two approaches that I believe fit the bill, approaches I call "interpretive" and "clinical" climate ethics. Neither is open to any of the objections laid out in last section, and both, I believe, are more likely to influence the policy community than the utopian approach that I have criticized.

Interpretive Climate Ethics

One option for philosophers wanting to influence policy discussions and policymakers is to abandon the attempt to develop and advocate for particular proposals, aiming instead to clarify the moral stakes of policy disputes, which are often unclear both to participants in and observers of those disputes. More specifically, interpretive climate ethics seeks to engage with live policy debates by looking at the various positions being defended, identifying the ethical perspectives informing those positions, and offering some guidance for those attempting to navigate any relevant tradeoffs. Instead of advocating for any particular position, the goal of this approach is to help participants in the relevant debates think more clearly about the moral issues involved.

There are plenty of examples of this sort of work. One particularly notable example of this sort of work is chapter 3 of Working Group III's contribution to the 2014 IPCC Assessment Report, which seeks "to provide a framework for viewing and understanding the human (social) perspective on climate change, focusing on ethics and economics" (Working Group III 2014, 213). Another is David Morrow's recent book, *Values in Climate Policy* (Morrow 2020), which identifies and clarifies the moral viewpoints implicit in a wide variety of climate policy proposals. Yet another, particularly timely example—this one unrelated to climate change—is the framework developed by a group of bioethicists to guide policymakers' thinking about ethical issues related to reopening the economy after COVID-19-related shutdowns (Bernstein et al. 2020). Even so, if my own experience is any indication, there is a real need for more of this sort of work, as it seems to me relatively rare in policy discussions for ethical issues to be understood as well as, say, economic ones.

In addition to helping to clarify the moral stakes of policy disputes, interpretive climate ethics has the potential to help non-philosophers understand philosophy's relevance to their own work. By doing more interpretive work, therefore, philosophers could not only increase the impact of their work on policy, but also improve the discipline's standing among policy professionals, who sometimes struggle to see what the discipline has to offer them.

Clinical Climate Ethics

I think there is a lot to be said for the interpretive approach, but many climate ethicists will be more interested in assessing and developing policy recommendations themselves than in helping guide policymakers' own thinking about them. My recommendation to them is to adopt an approach to non-ideal theory first elaborated by David Wiens (2012). In light of its emphasis on diagnosing political problems and prescribing remedies for them, I'll call this approach "clinical climate ethics."[14]

The clinical climate ethicist proceeds in three phases. The first is the identification phase, in which the clinician seeks to ensure she has identified a problem that both can and should be solved. The rationale for this step is simple. If there are not in fact any feasible, morally superior alternatives to the problematic policy or institution, then it would seem that the relevant policy or institution cannot be improved, at least for the time being.[15] In such cases, as unfortunate or harmful as the current state of things may be, there will be little hope of persuading policymakers to try to change it.[16] If, however, she *can* identify feasible alternatives to the apparently problematic policies or institutions in question that are more just than the present arrangements, she can credibly claim to have identified a "remediable injustice."[17]

To do so, the clinician first compares some apparently problematic policy or institution, not to some purportedly ideal social world but to feasible alternative policies and institutions. In some cases, these alternative policies and institutions may be counterfactual. Caution is in order, however, as it can be hard to know whether or not it would be feasible to implement counterfactual policies or institutions; just consider, for example, how unlikely it would have seemed in January 2020 that entire countries would shut down much of their economies, drastically reduce travel and in-person social interaction, transition to online schooling, and otherwise adapt to COVID-19 as they did just a few months later.[18] Consequently, the safest way for the clinician to be sure she is focusing only on feasible alternative social arrangements may be for her to compare the policies and institutions in question to those that have been implemented elsewhere. If, for example, a clinician believes that the top tax rates in America should be higher, she could begin by comparing them to the top tax rates in other countries to check whether there is reason to think that they even could be. (They could.) At the same time, though, it is important for clinicians not to unduly limit the contrast class, since this is liable to lead them to discount as insoluble problems that we might in fact be able to solve.

Once the clinician has identified a set of feasible alternatives, she sets about ordering them according to how just they are. At this stage, the ordering can be rough, since the point of this procedure is not to arrive at some definitive ordering but to arrive at some rough but defensible ordering that will allow her to justify her claim to have identified a remediable injustice. All she needs for that purpose is a rough, plausible ordering of nearby worlds in terms of justice that ranks the actual world below some alternative, feasible world.

Once some remediable injustice has been identified, the clinician proceeds to the second, diagnostic phase, during which she does her best to diagnose the problem she has identified. How, she asks, did the problem come about—what exactly went wrong, and why? Answering this question involves both moral and causal analyses.

The moral analysis is the aspect of clinical work that will be most familiar to philosophers and the aspect to which much existing work by ethicists will be most relevant. Its point is to determine what has gone wrong from a moral point of view—to get clear, as Wiens puts it, about "the ways in which a social process or outcome constitutes a *moral* problem" (Wiens 2012, 61). The clinician may consider whether the policy or institution fails to appropriately embody some important ideal, or more broadly, why that ideal is important in the first place, and what it would take to better realize it. For instance, this part of the clinician's work might involve considering why exactly, and in what sense, equality is worth caring about (Anderson 1999). It might also involve thinking about what rights we take people to have, as well as about the conditions in which we take those rights to have been violated—whether, for instance, people have a right to a stable climate system and what it would mean for that right to be upheld. By helping us to understand what has gone wrong, morally speaking, this sort of analysis can help us get clearer about the nature of the failure the clinician seeks to address and what it would take to address it.

The point of the causal analysis, in contrast, is to identify and understand the relationships between the various causal processes that produce the problematic outcome(s). As Wiens stresses, this process necessarily involves appropriate engagement with relevant literature in the social sciences (Wiens 2012, 63). I would add that, in some cases, it may also involve engaging with literature in the natural sciences, especially when it comes to environmental problems like climate change.

Finally, the clinician proceeds to the design phase, where she develops and proposes a remedy for the problem she has identified and diagnosed. In doing so, she draws on the results of the second phase to identify a set of feasible solutions that lack the moral problems of the policies or institutions she means to replace or improve, while being sure not to replicate the causal processes that produced the result that was the impetus for her clinical work. In addition, she aims to ensure that the policies or institutions she recommends will avoid not only the failings she has identified in the present policies or institutions but also any other *potential* failings (cf. Wiens 2012, 65). Once she has identified candidate solutions, the clinician recommends the best one, that is, the one that best addresses the moral failings identified in the diagnosis. Wiens does not consider how the clinician ought to choose between multiple feasible solutions that avoid both the moral failings of the problematic conditions and the causal processes responsible for them. I would suggest that, in such cases, the clinician must do her best to balance the moral desirability of candidate solutions against their likelihood of being implemented and withstanding whatever pressures it may face thereafter.

It is worth noting that the scope of the feasibility assessments relevant for the design phase is much narrower than for those relevant to the identification phase. This is because feasibility assessments are always relative to the agent who would perform the act in question and the period during which they would perform it, and the relevant agents and timeframes differ in the two stages.[19] When, in the identification phase, we are trying to determine whether it is so much as possible for societies like our own to solve the problem in question, the relevant agents are any societies sufficiently similar to ours, and the relevant timeframes are the entirety of those societies' existence. (The metric of sufficiency here will be whether the societies in question differ so much from ours that policymakers will think it irrelevant what they were able to accomplish.) In contrast, during the design phase, when we are trying to determine which among the possible solutions is feasible so as to make recommendations for action in the near term, the relevant actor will be whichever legislative body, government official, or regulatory body that has the power to enact the proposed reform, and the relevant timeframe will be the next several months or, at most, a year or two.[20]

As the foregoing may already have made clear, the crucial difference between the clinical and utopian approaches is that, whereas the utopian approach is prospective, in the sense that it orients itself by considering what policies and institutions would be in place in some possible future ideal social order, the design phase of the clinical approach is, in Wiens' words, "largely retrospective in the sense that our sights are set by looking backward, at the places we've been rather than at the places we'd like to go to" (Wiens 2012, 66). In other words, "We design institutions to avert failure, not to realize an ideal" (ibid.). Consequently, the clinical approach does not at any point require us to consider purportedly ideal social orders with no regard for the feasibility of realizing them.

Some readers might worry that this cannot be right, that in fact clinical work *must* be prospective in this sense. But this is just not true. Rather, as Wiens notes, ordinary moral reasoning is perfectly sufficient. To determine which of the solutions in the feasible set is best, we need only

> take our current social conditions and compare them to the conditions that would arise were we to implement some particular feasible institutional solution. We then ask ourselves: Do we think the counterfactual conditions are acceptable? Are the counterfactual conditions an improvement upon current conditions? On the basis of which principles do we make these judgments? Can we justifiably endorse these principles upon reflection? (Wiens 2012, 67)

So long as we can do this, we do not need to know either what the ideal social order would be like or how various solutions compare to it.

Two Objections to the Clinical Approach

The clinical approach is open to at least two objections. The first objection was raised by A. John Simmons against Amartya Sen, whose own recommended approach to non-ideal theory is similar to the clinical approach. Sen argues that identification of the ideal is neither necessary nor sufficient for determining which of two alternatives is more just and suggests that, if we want to do that, we should instead simply compare them directly (Sen 2009, ch. 4). He illustrates the necessity point with an analogy:

> We may be willing to accept, with great certainty, that Mount Everest is the tallest mountain in the world, completely unbeatable in terms of stature by any other peak, but that understanding is neither needed, nor particularly helpful, in comparing the peak heights of, say, Mount Kilimanjaro and Mount McKinley. (Sen 2009, 102)

He continues: "There would be something deeply odd in a general belief that a comparison of any two alternatives cannot be sensibly made without a prior identification of a supreme alternative" (ibid.). By way of response, Simmons argues that Sen's point about comparing mountains does not have the methodological significance he takes it to have. He writes:

> which of two smaller "peaks" of justice is the higher (or more just) is a judgment that matters conclusively only if they are both on equally feasible paths to the highest peak of perfect justice. And in order to endorse a route to that highest peak, we certainly do need to know which one that highest peak is. Perhaps for a while we can just aim ourselves in the general direction of the Himalayas, adjusting our paths more finely—between Everest and K2, say—only when we arrive in India. But we need to know a great deal about where to find the serious candidates for the highest peak before we can endorse any path to them from here. (Simmons 2010, 35)

Because it ignores this crucial point, the thought goes, Sen's comparative approach is liable to lead us away from the ideal, even if it reliably steers us toward the most just option available to us. Similarly, someone might suggest, for the clinical approach.

In response, I would begin by noting that, given the problems with the utopian approach I outlined earlier, it seems to me an open question whether the ideal targets approach, which Simmons seems to have in mind, is any less likely to lead us away from the ideal (cf. Wiens 2015b, 472). After all, that approach is liable to suggest we take the initial step toward realizing purportedly ideal worlds even though we will often struggle to determine both whether we will be able to take subsequent ones and whether, if we are and we take them, we will be glad we did. Perhaps the best we can hope for is something like the vision outlined by Gaus, wherein diverse communities conduct Millian experiments in living and share insights about the solutions that have worked for them, slowly but surely getting a better sense for what ways of organizing societies will truly make things better (Gaus 2016, esp. §4.3). Whether or not that is right is a fascinating question that is, unfortunately, far beyond the scope of this chapter. Even if identifying the ideal were necessary to determining what we ought to do here and now, whether or not that fact vindicated the ideal targets approach as a way of developing and arguing for policy recommendations against the practical objections I have raised would depend on the attitudes of relevant policymakers. If policymakers did not find persuasive the sorts of arguments that approach suggests, it would remain true that making those arguments is not a good strategy for philosophers seeking to influence policy. In that case, the right approach for such philosophers might be to use the ideal targets approach to determine which policies are best but, when addressing the policy community, argue for policy proposals not by emphasizing their potential role in helping us to realize the ideal, but by arguing that the near-term benefits of adopting them are greater than those of relevant alternatives.

The second objection to the clinical approach is that it is too conservative, in the sense that it is liable to lead to complacency regarding prevailing societal injustices (cf. Wiens 2012, note 41). By requiring us to restrict our attention to feasible solutions to discrete problems with the present order, it may keep us from rectifying injustices whose causes are deeper and so cannot be addressed without far-reaching changes to the status quo that may not be feasible right now.[21] In this sense, the clinical approach is supposed to contrast with the utopian approach I have criticized. The thought is that, because the latter begins by considering the shape of utopia and proceeds to compare the actual world to the ideal, the utopian theorist's sense of what the problems and solutions are will be less constrained by the way things are.

If it were the case that the clinical theorist could only recommend currently feasible policies or institutional reforms, it would be too accommodating to current political realities. But there is no reason why the clinical theorist should be so limited. For though Wiens does not consider this possibility himself, it would seem that, in principle, there is nothing stopping the clinical theorist from recommending that relevant actors seek to eliminate some feasibility constraint that currently rules out some potential solution that, were it only feasible, would be superior to any currently available. For example, at present the filibuster is thought by many to be among the most significant obstacles to the enactment of ambitious climate policy in the United States. Should she come to believe this is correct, there is no reason why a clinician concerned about the United States' greenhouse gas emissions could not recommend that Senators change Senate rules to eliminate the filibuster as soon as possible.

With that said, though, I want to stress that clinicians recommending the elimination of some feasibility constraint ought to bear in mind the difficulties about assessing the character of and feasibility of bringing about far-off worlds that I pressed earlier. The further away are the worlds in which the relevant feasibility constraint is eliminated, the harder it will be to know whether or not the clinician's preferred solution will indeed become feasible once the relevant feasibility constraint is eliminated, and the less persuasive policymakers will tend to find the clinician's argument for her proffered remedy. Of course, there will sometimes be cases in which some clinician feels that, because nothing less would adequately address the injustice she has identified, she cannot in good conscience recommend anything short of a very radical solution that would require the elimination of multiple feasibility constraints. In cases like this, I would suggest that clinicians are more likely to find an audience with NGOs and activists, who do not face the same constraints and are not subject to the same pressures as politicians and, as a result, are often better positioned to undertake these sorts of political projects.[22]

I therefore doubt that the clinical approach is in the relevant sense too conservative. However, I hasten to add that the objection is right to suggest that the risk of a kind of status quo bias is real. Biases acquired during an upbringing in a society that is deeply unjust in many ways may lead clinical theorists to dismiss, discount, or pay insufficient attention to members of marginalized or oppressed groups when they speak about the problems they face. As a result, clinicians may fail to notice or comprehend or take sufficiently seriously what are in fact pressing social problems. In addition, they may dismiss some candidate solution as infeasible because they think it is unlikely to be sufficiently palatable to legislators or regulators when, in fact, it could easily gain traction.

Charles Mills has argued that ideal theorists are also at risk of making these sorts of errors (Mills 2005, 2007, 2018). In fact, I would go further and say that anyone working to find solutions to societal problems runs this risk. If the clinical approach has any advantage over others on this front, it is that it requires clinical theorists to diagnose the failures on which they focus and to determine whether or not candidate solutions have the same features that produced the failure in the first place. It thereby forces clinical theorists to confront the nitty-gritty realities of the problems they want to solve rather than allowing them to stay at the level of abstract theory. This sort of acquaintance with the realities of social ills can make it harder to maintain mistaken preconceptions and can open one's mind to other ways of looking at things, especially if it involves direct interaction with the people harmed by the present order (cf. Wiens 2012, 62; Mills 2005).

How to Approach Interpretive and Clinical Work

I conclude this discussion with three points about how philosophers who want to maximize their influence on policy should go about doing the type of work I have recommended. First, philosophers' work is unlikely to influence policy unless they present it in such a way as to maximize the likelihood that it will reach decision-makers and relevant experts, and to do that, they will need to consider different venues and media types. Though they have their place, dense, pay-walled journal articles and books are far less likely to reach key audiences than are articles on popular websites like *Vox* or *Politico*. Podcasts, YouTube videos, and infographics are excellent choices as well. Ethicists should also try to time interventions so as to maximize relevance. For instance, advocacy of ambitious domestic emissions reduction measures is much less likely to find an audience among policymakers when the party in power denies the reality of climate change.

In addition, the best interpretive and clinical work will be developed in close coordination with the people who would be affected by the policies under discussion, relevant policy experts, and, where possible, policymakers themselves, as this will help to ensure that philosophers' recommendations are responsive to relevant concerns and are appropriately sensitive to the constraints policymakers face.[23] This sort of engagement could come through participation in formal committees or working groups, or through conducting interviews, but it could also take forms less unusual in philosophy, such as co-authoring pieces.

Finally, the best interpretive and theoretical work will eschew the tendency of much non-ideal theory to vaguely discuss what some impersonal "we" should do and instead take care to identify and direct itself to actors who are both willing and able to affect any relevant reforms (cf. Laurence 2020). In addition to helping to make clearer the practical implications of philosophers' work, getting clear about the relevant agents of change will also inform the content of clinical work, helping to ensure that any recommendations or guidance offered by philosophers are actually useful to those in a position to advance climate action.

III. CONCLUSION

I began by criticizing the utopian approach to climate ethics, suggesting that the arguments for near-term reforms it suggests are liable to be unconvincing to policymakers. In addition, I noted that those who draw on the ideal benchmarks approach commit the fallacy of approximation. I then went on to describe and defend two alternative ways of thinking philosophically about climate policy, interpretive and clinical climate ethics, neither of which is open to the objections I leveled at the utopian approach. In closing, I want to say a bit about who this essay is for.

It has been no part of my intention here to deny that there are lots of other, perfectly acceptable ways philosophers might wish to engage with political discussions or try to advance climate action. For instance, not all philosophers will be concerned about the extent to which their work will influence policy outcomes. Some will agree with Gerry Cohen that the role of political philosophy is, not to make concrete practical recommendations for action, but to discern the character of justice more abstractly conceived and will therefore focus on that sort of work (Cohen 2003). Even among those philosophers who do want their work to influence policy outcomes, many will be content to influence policy outcomes indirectly. These philosophers might choose to continue exploring ideal worlds in an attempt to identify barriers to their realization, break down the cognitive and attitudinal feasibility constraints that prevent it, or inspire activists. Alternatively, they might focus their efforts on the sort of conceptual innovation discussed by Raymond Geuss, aiming to work out new concepts and ways of thinking so as to enable people to give voice to and understand phenomena that are difficult either to describe or make sense of (Geuss 2008, 42–50). Miranda Fricker's discussion of the origin of the concept of sexual harassment illustrates the potential and power of this sort of work (Fricker 2007, §7.1). I am not suggesting that there is anything amiss in these kinds of projects; nor, however, are the philosophers who prefer them to the sort of work I have recommended its intended audience.

This essay is aimed at those philosophers who wish to influence climate policy more directly, for instance by influencing policy design, by convincing policymakers to support a particular policy change or institutional reform, or perhaps by suggesting longer-term projects for NGOs and activists that would help to expand the bounds of the politically possible. I have argued that philosophers who want to advance climate justice in this way would do well to adopt the interpretive and clinical approaches I have outlined. As they do so, they should seek to ensure that their work is timely, that it is presented in ways and in venues that maximize its chances of reaching key audiences, that their recommendations are targeted toward key actors, and that they carry out their work in conjunction with relevant practitioners where possible. To these philosophers, I say: Welcome! Let's get to work.[24]

NOTES

1. The utopian approach I criticize here is identical to what David Wiens has called "the ideal guidance approach," but my criticisms of this view differ from those he develops in his 2015a.

2. For relevant passages in Rawls, see Valentini (2012, §3). Recent defenses of this approach include Simmons (2010) and Valentini (2009).

3. Quite possibly this is true more generally of philosophy's influence on the policy community, but because my experience and expertise is limited to climate policy in particular, this amounts to little more than speculation on my part.

4. The most explicit expression of this view I have been able to find is at Valentini (2009, 333), but it seems safe to say it is fairly widespread.

5. I adopt this terminology from Wiens (2015a).

6. On the benefits of natural climate solutions such as preserving and restoring tropical forests, see Griscom et al. (2017).

7. Cf. Gaus' discussion of what he calls "the choice" at Gaus (2016, 82–84, 142–144).

8. In fairness to Wainwright and Mann, policymakers are clearly not their target audience. Instead, they seem to understand their work as a sort of guide for activists (see especially their ch. 7). So understood, Wainwright and Mann's project makes more sense, since it could very well prompt thinking about alternatives to the present social order—one of several roles for which utopian thinking seems well suited, as I say later. Even so, the lack of detail in their description of Climate X seems likely to lessen usefulness for activists, as well as its impact.

9. The language of "hard" and "soft" feasibility constraints comes from Gilabert and Lawford-Smith (2012).

10. I regard the Mary Robinson Foundation's "Global Guardians" idea as a much more realistic proposal in the same vein. See Mary Robinson Foundation (2018).

11. Some readers will wonder whether this line of thought depends on a particular conception of feasibility. Does it matter, for instance, whether feasibility is better understood as a categorical matter (as Wiens' discussion suggests) than as a scalar one (as Gilabert and Lawford-Smith suggest)? But, at least for the argument in this section, I really don't think it does. Any plausible conception of feasibility is going to entail that feasibility assessments consider a very large number of factors that will often be difficult to gauge and that are subject to change over time in ways that are hard to predict. So long as that's the case, it will be reasonable for policymakers to worry that, even if subsequent steps on the path to realizing the ideal seem feasible now, we could very well be mistaken, and even if not, they might no longer be feasible when we get there. That is all my argument here requires. (Thanks to the editors for pressing me to consider this point.)

12. For additional discussion of Abizadeh's argument, see Miller (2010) and Abizadeh (2010).

13. Many thanks to my good friend Adam Reid for pressing me on this point.

14. In my presentation of the clinical approach, I follow Wiens in taking for granted his own, categorical conception of feasibility. While I think it is probably possible to describe the clinical approach or something very close to it in a way that instead draws on other conceptions of feasibility, some of the details would have to be significantly different. Rhetorical expediency therefore demanded that I pick one conception. Since Wiens himself presupposes his categorical conception of feasibility in his exposition of the clinical approach, and since I personally prefer Wiens' conception to its main competitor in the literature, Gilabert and Lawford-Smith's, I have chosen his.

15. Here I'm reminded of Churchill's well-known remark about democracy: it's the worst form of government, aside from all the others.

16. This political twist on the identification phase is my own addition to Wiens' proposed approach.

17. The term "remediable injustice" comes from Sen (2009, vii).

18. My thanks to Sarah Kenehan for emphasizing the significance of this example.

19. Wiens explains how this point fits into his model at Wiens (2015b, 459–460).

20. This is my own addition. Wiens does not suggest that the scope of the feasibility assessments relevant for each of the two stages should be different.

21. Compare Gilabert and Lawford-Smith's concern to avoid what they call "cynical realism" at Gilabert and Lawford-Smith (2012, 815).

22. While the basic idea is the same in each case, it is worth noting that, if the idea is to develop recommendations for NGOs or activists, the clinical approach would have to be slightly different from that outlined earlier. In particular, the scope of the feasibility assessments relevant for the first and third steps would have to be different. In this case, the metric of sufficiency relevant for the identification phase here will be whether the societies in question differ so much from ours that NGOs or activists will think it irrelevant what they were able to accomplish, and the relevant agents for the design phase will be NGOs or activists. In addition, the relevant timeframe may be a bit longer than that relevant for policymakers.

23. Such work would be "engaged" in the sense recently elaborated by Fergus Green and Eric Brandstedt (2020), whose discussion makes a number of excellent, complementary points about how philosophy can best advance climate action.

24. Opinions expressed here are my own and do not reflect those of Climate Advisers. I am indebted to Sarah Kenehan, Corey Katz, Adam Reid, Angela Hvitved, and Damian Dalle Nogare for invaluable comments on earlier drafts. The chapter is far better for their input. In addition, I am grateful to my wife, Erika, who cared for our infant daughter during the many nights and weekends I spent researching and writing it. Without her help, there is no way I could have completed this project.

REFERENCES

Abizadeh, Arash. 2008. "Democratic Theory and Border Coercion." *Political Theory* 36(1): 37–65.

Abizadeh, Arash. 2010. "Democratic Legitimacy and State Coercion: A Reply to David Miller." *Political Theory* 38(1): 121–130.

Anderson, Elizabeth. 1999. "What is the Point of Equality?" *Ethics* 109(2): 287–337.

Bernstein, Justin, Brian Hutler, Travis N. Rieder, Ruth Faden, Hahrie Han, and Anne Barnhill. 2020. "An Ethics Framework for the COVID-19 Reopening Process." https://bioethics.jhu.edu/research-and-outreach/projects/grappling-with-the-ethics-of-social-distancing/.

Cohen, Gerry A. 2003. "Facts and Principles." *Philosophy & Public Affairs* 31(3): 211–245.

Estlund, David. 2020. *Utopophobia*. Princeton: Princeton University Press.

Fricker, Miranda. 2007. *Epistemic Injustice: Power and the Ethics of Knowing*. Oxford: Oxford University Press.
Gardiner, Stephen M. 2014. "A Call for a Global Constitutional Convention Focused on Future Generations." *Ethics & International Affairs* 28(3): 299–315.
Gardiner, Stephen M., and David A. Weisbach. 2016. *Debating Climate Ethics*. New York: Oxford University Press.
Gaus, Gerald. 2016. *The Tyranny of the Ideal: Justice in a Diverse Society*. Princeton: Princeton University Press.
Geuss, Raymond. 2008. *Philosophy and Real Politics*. Princeton: Princeton University Press.
Gilabert, Pablo, and Holly Lawford-Smith. 2012. "Political Feasibility: A Conceptual Exploration." *Political Studies* 60: 809–825.
Green, Fergus, and Eric Brandstedt. 2020. "Engaged Climate Ethics." *The Journal of Political Philosophy*. Advance online publication. https://doi.org/10.1111/jopp.12237.
Griscom, Bronson W., Justin Adams, Peter W. Ellis, Richard A. Houghton, Guy Lomax, Daniela A. Miteva, William H. Schlesinger, David Shoch, Juha V. Siikamäki, Pete Smith, Peter Woodbury, Chris Zganjar, Allen Blackman, João Campari, Richard T. Conant, Christopher Delgado, Patricia Elias, Trisha Gopalakrishna, Marisa R. Hamsik, Mario Herrero, Joseph Kiesecker, Emily Landis, Lars Laestadius, Sara M. Leavitt, Susan Minnemeyer, Stephen Polasky, Peter Potapov, Francis E. Putz, Jonathan Sanderman, Marcel Silvius, Eva Wollenberg, and Joseph Fargione. 2017. "Natural Climate Solutions." *PNAS* 114(44): 11645–11650. https://www.pnas.org/content/114/44/11645
Laurence, Ben. 2020. "The Question of the Agent of Change." *The Journal of Political Philosophy* 28(4): 355–377.
Mary Robinson Foundation. 2018. "Global Guardians: A Voice for Future Generations." https://www.mrfcj.org/wp-content/uploads/2018/02/Global-Guardians-A-Voice-for-Future-Generations-Position-Paper-2018.pdf.
Miller, David. 2010. "Why Immigration Controls Are Not Coercive: A Reply to Arash Abizadeh." *Political Theory* 38(1): 111–120.
Mills, Charles. 2005. "'Ideal Theory' as Ideology." *Hypatia* 20(3): 165–184.
Mills, Charles. 2007. "White Ignorance." In *Race and Epistemologies of Ignorance*, edited by Shannon Sullivan and Nancy Tuana, 13–38. Albany: SUNY Press.
Mills, Charles. 2018. "Through a Glass Whitely: Ideal Theory as Epistemic Injustice." *Proceedings and Addresses of the APA* 92: 43–77.
Morrow, David. 2020. *Values in Climate Policy*. London: Rowman & Littlefield.
Sen, Amartya. 2009. *The Idea of Justice*. Cambridge: Harvard University Press.
Simmons, A. John. 2010. "Ideal and Nonideal Theory." *Philosophy & Public Affairs* 38(1): 5–36.
United Nations Environment Programme (UNEP). 2018. *The Adaptation Gap Report 2018*. Nairobi, Kenya: UNEP. https://www.unenvironment.org/gan/resources/publication/adaptation-gap-report-2018.
Valentini, Laura. 2009. "On the Apparent Paradox of Ideal Theory." *The Journal of Political Philosophy* 17(3): 332–355.

Valentini, Laura. 2012. "Ideal vs. Non-ideal Theory: A Conceptual Map." *Philosophy Compass* 7(9): 654–664.
Wainwright, Joel, and Geoff Mann. 2018. *Climate Leviathan: A Political Theory of Our Planetary Future*. London: Verso.
Weins, David. 2012. "Prescribing Institutions Without Ideal Theory." *The Journal of Political Philosophy* 20(1): 45–70.
Weins, David. 2015a. "Against Ideal Guidance." *The Journal of Politics* 77(2): 433–446.
Weins, David. 2015b. "Political Ideals and the Feasibility Frontier." *Economics and Philosophy* 31(3): 447–477.
Working Group III to the Fifth Assessment Report of the Intergovernmental Panel on Climate Change. 2014. *Climate Change 2014: Mitigation of Climate Change*. Cambridge: Cambridge University Press.

Chapter 3

Falling on Your Own Feasibility Sword? Challenges for Climate Policy Based on "Simple Self-Interest"

Stephen Gardiner and Justin Lawson

Some insist that ethics is central to international climate policy. Others, who we call the "economic realists," argue that it is largely irrelevant, and that talk of climate justice in particular should be excluded. Instead, they insist, climate policy must be pursued solely in terms of self-interest, because any other approach is not "politically feasible."[1] The negative arguments such authors put forward against ethics and justice have received considerable attention.[2] However, so far less has been said about their positive proposals. In this chapter, we address this gap by examining a specific argument for self-interested climate policy offered by one of its most prominent proponents, David Weisbach.

Weisbach aims to show that "simple self-interest" (Gardiner and Weisbach 2016, 170) justifies "aggressive" emissions reduction policies "far more ambitious than those currently on the table" (ibid, 146). He argues that this is true under "almost any plausible assumptions" about climate change and "robust to a broad range" of scientific and economic assumptions (ibid, 170). Such claims are initially appealing to climate advocates desperate for real progress, including those who embrace ethics. Thus, it is helpful to see a detailed defense in print.

Nonetheless, sadly, we shall conclude that Weisbach's positive argument does not succeed. The initial rhetoric overstates the actual results, and the argument itself suffers from serious problems concerning ambition, its understanding of self-interest, and its basic scientific methodology. One particular concern is that the self-interest approach may "fall on its own sword." The positive argument is held hostage to the same kinds of worries about political feasibility that economic realists deploy against ethics- and justice-based approaches. Indeed, Weisbach insists on *such a tight set* of feasibility constraints that his argument calls into question the *very possibility* of appropriately ambitious climate action. This provides another (indirect) argument against economic realism: it is not "what works."

We conclude that the case for climate policy based solely on self-interest remains underdeveloped, inconclusive, and ultimately misleading. Moreover, the gaps suggest that what is needed instead is an approach that embraces ethics and justice. Thus, the failure of Weisbach's positive argument has wider implications. It reveals challenges that any self-interest approach to climate policy must address.

I. CONTEXT

In *Debating Climate Ethics*, David Weisbach offers a positive argument for climate action based on self-interest. He describes the self-interest approach as "the major alternative" to justice and ethics (ibid, 170). He also stresses that his account is based on "simple self-interest" (ibid, 170). It tells us "what we should do about climate change if we act purely to save our own necks" (ibid, 170) or how to behave if we have "a desire to stop hitting ourselves in the head with a hammer" (ibid, 197).

Weisbach's Claims

Weisbach makes impressive claims on behalf of his positive argument. First, he says that it delivers dramatic conclusions:

> "following our *self-interest* will . . . lead . . . to the *requirement* of *drastic emissions reductions* and an energy transition *as soon as possible*" (ibid, abstract);

> "*self-interest* . . . *demands* that we pursue *aggressive* policies to reduce emissions, policies that are *far more ambitious than those currently on the table*" (ibid, 146);

> "We *must* reduce emissions to zero in the *near future*" (ibid, 176);

> "We *need* to reduce emissions *rapidly*." (ibid, 146)[3]

In summary, "simple self-interest" "requires" and "demands" "aggressive" policies, "far more ambitious than those currently on the table," involving "rapid" and "drastic" emissions reductions, "an energy transition as soon as possible," and "zero or near-zero" emissions "in the near future" (ibid, abstract, 146, 176).

Second, Weisbach claims that the argument from simple self-interest succeeds under "almost any plausible . . . scientific and economic assumptions":

> under *almost any plausible assumptions* about climate change, it is in our self-interest to start reducing emissions now, on a global basis, and to reduce emissions to near zero in the not-too-distant future. This conclusion is *robust to a broad range of assumptions* about the science and the economics of climate change. (ibid, 170)

In essence, the claim is that the positive argument should be uncontroversial, since it does not depend on contentious scientific or economic claims, and so covers all relevant scenarios.

Third, Weisbach asserts that the positive argument achieves all of this solely on the basis of "simple," "narrow," and uncontroversial assumptions about self-interest:

> "even using *narrow notions of self-interest*, it is in our self-interest . . . *to reduce emissions, starting now, and to do so rapidly*" (ibid, 149);

> "*policies based purely on self-interest*—a desire to stop *hitting ourselves in the head with a hammer*— . . . take us to *aggressive* reductions in emissions" (ibid, 197);

> "It is in our self-interest to . . . reduce emissions to near zero in the not-too-distant future" (ibid, 170);

> "*solely* [on the basis of] the *simple self-interest* of people who are alive today, their children, and their grandchildren." (ibid, 170)

> "*relatively narrow notions [of self-interest] are sufficient to motivate action on climate change.* Not limiting climate change will directly hurt us, our children, and our grandchildren in straightforward ways. Climate change threatens our food supplies, our cities through storms and sea level rise, and, generally, our lives as we have come to know them. While I believe in a broad notion of self-interest and well-being, *we need not engage in debates about exactly what this means* to know that we should want to limit climate change." (ibid, 154)

This is exciting stuff. If successful, Weisbach would have provided a strong argument for climate action that is independent of ethics and justice. Even those who are skeptical of his objections to ethics may be glad to see such an argument, since it might complement and reinforce ethics-based arguments for climate action. If a simple, positive argument from self-interest succeeds, it raises the hope that many obstacles to robust climate action might be overcome or deferred, at least for a time (cf. Gardiner 2011).

Our Objections

Sadly, we believe that Weisbach's positive argument fails. We shall offer four, overlapping reasons for this failure.

The first is that there is a serious mismatch between the initial rhetoric and the reality of what the actual argument is able to deliver. Specifically, rather than a robust, ambitious case for climate action, Weisbach's positive argument yields only relatively weak and tentative conclusions. Moreover, his approach is, if anything, notably less aggressive than more familiar proposals.

The second reason is that the discussion of self-interest is unhelpful. For one thing, the argument appears to oscillate between distinct and indeed rival accounts of self-interest, some of which may be "narrow" and "simple," but others of which are clearly not. This makes it challenging to maintain a consistent interpretation of Weisbach's argument, let alone one that would justify his "ambitious" conclusions. For another thing, the account of self-interest that ultimately drives Weisbach's central feasibility constraint is a short-term, narrowly economic one. In our view, this constraint is implausible, and does not take seriously the challenge of our historical moment.

The third reason for failure is that, because of the impact of the short-term, narrowly economic feasibility constraint, Weisbach's positive argument appears to fall afoul of the same kinds of worries about political feasibility that he himself deploys against ethics- and justice-based approaches. Indeed, Weisbach insists on *such a tight set* of feasibility constraints that his position actually calls into question the *very possibility* of appropriately ambitious climate action. This is a challenge that Weisbach himself must confront; as he puts it, "feasibility concerns run both ways" (ibid, 188).

The fourth reason that the simple self-interest argument fails is that some of Weisbach's claims rest on a dubious scientific methodology. For instance, Weisbach infers his feasibility constraints from past empirical regularities in ways that are highly questionable. In doing so, he risks mistaking core causes of the problem for constraints on solutions. Thus, economic realism threatens to hide indefensible behavior behind the pseudo-scientific language of "policy constraints."[4]

Preview

To get a sense of where we will be going, let us offer a preview of how we understand part of Weisbach's argument, and in particular the disparity between rhetoric and results. Weisbach summarizes the main claims of his argument under the following headings:[5]

C1: Emissions must go to zero
C2: Emissions must go to zero soon
C3: We need to start reducing emissions now
C4: Everyone must start now
C5: Uncertainty strengthens these conclusions

We will focus on the second, third, and fourth claims. In particular, we shall argue that a more appropriate summary of the actual arguments in Weisbach's chapter would be:

C2*: Emissions must go to zero sometime between 2020 and 2100, or possibly by 2150.
C3*: We should start now and proceed slowly. We should not build any new fossil fuel power plants. But it is not politically feasible to replace existing infrastructure before the end of its normal working life of 50–100 years. Some new fossil fuel plants are inevitable too.
C4*: It is incompatible with climate goals for some countries, like China and India, to continue to build new coal-fired power plants at their current pace. However, we cannot expect them not to build more if they perceive this as necessary for their economic growth.

We submit that these are dramatically different claims. In the remainder of the chapter, we shall show how they emerge from our reading of Weisbach. We divide our discussion into three main sections: on ambition, self-interest, and methodology.

II. AMBITION

Zero Soon

Let us begin by examining Weisbach's second claim (C2), that emissions must go to zero soon.

To understand (C2), we must first say something about (C1), that emissions must go to zero eventually. To establish (C1), Weisbach argues that once released into the atmosphere carbon dioxide is effectively permanent, that temperatures will continue to go up when we emit more, and that there is a limit to tolerable temperature increases. On the basis of these claims, he concludes, "we eventually have to reduce emissions to zero or near zero" (ibid, 176). He goes on to say, "this is true even if the temperature turns out to be relatively insensitive to greenhouse gases and even if the harms turn out to be on the lower end of the possibilities" (ibid, 176-7).

The key idea behind (C1) appears to be that there is a limit to tolerable temperature increases, and so also to greenhouse gas emissions. Given this, the operative question is how many emissions are tolerable, and when that tolerable limit will be reached. In mainstream climate policy, this idea is often represented in terms of a global carbon budget. Many have argued that we should be thinking of climate mitigation in terms of "the cumulative amount of 'allowable' carbon emissions to meet a global temperature target" (Lahn 2020), or, as we prefer to put it, the amount of emissions available before humanity runs an unacceptable risk of breaching key climate targets.[6]

As we have indicated, much of this approach is common in climate circles. We would quibble with some aspects of it (see below) and also with Weisbach's ability to endorse it based on simple self-interest.[7] Nevertheless, for now, let us move on to (C2) itself, the claim that emissions must go to zero *soon*, as this is closer to the heart of Weisbach's argument.

Recall that Weisbach asserts that his argument justifies "aggressive" policies, "far more ambitious than those currently on the table," where this requires "rapid" and "drastic" emissions reduction, "an energy transition as soon as possible," and "zero or near-zero" emissions "in the near future." How well does this rhetoric match his results?

Timeframe

Let us begin with the *timeframe*. What does Weisbach mean by "soon"? The answer is surprising: "We have to reduce emissions dramatically, to near zero, *by sometime around the end of this century*" (Gardiner and Weisbach 2016, 179, emphasis added); "the right target date is somewhere in this range . . . we need to be thinking about *a 100-year horizon*" (ibid 179, emphasis added). These conclusions are drawn from a very rough sensitivity analysis. In order to limit the global temperature rise to between 2 and 4°C, Weisbach argues that emissions must go to zero sometime between now (2020) and the end of the century (2100), or perhaps even the middle of the next century (2150).

One problem surrounds the rhetoric. Arguably, it is seriously misleading to frame "sometime around the end of this century" as "soon" or "in the near future"; it is at least very odd. Suppose we ask, "Soon, relative to what?" The natural answers might be: (i) soon relative to commonsense timeframes, or (ii) soon relative to standard timeframes in climate circles. But Weisbach's use of "soon" fits neither. It would be a more accurate summary of his position to say "zero sometime in our lifetimes, or those of our children (or perhaps grandchildren)," or even "zero sometime in living memory." Yet such summaries, while accurate, would likely sound at least complacent, and probably alarming to many concerned about climate. In light of this, Weisbach's emphasis on "soon" gives his proposal an *appearance* of urgency that it has not yet earned.

Why might this dispute about wording matter? One reason is that it is inaccurate for Weisbach to suggest that a policy that puts forth such a range ("a 100-year horizon") is *"far more ambitious"* than others currently on the table. By comparison, common benchmarks elsewhere in climate policy circles are proposals for net zero by 2050 or 2060, as endorsed by the IPCC, the EU, the UK, Japan, South Korea, and recently China.[8] Moreover, there are more radical proposals, such as those of the activist movement Extinction Rebellion, which calls for net zero by 2025. Given this, it is misleading to claim that Weisbach's position is much more aggressive than anything else out there. His proposal for a 100-year horizon is, if anything, markedly less ambitious than others, including the mainstream.

Target

Let us turn now to the question of the *target*.[9] Weisbach's case initially looks better if we turn his framing from one of timing ("soon") to one of *limits*. As we have seen, the first stage of Weisbach's positive argument is embedded in the familiar approach of proposing a global emissions budget. On this view, timing is not itself crucial. What is important is respecting a limit on total emissions; when they occur might not matter so much.

Probably the most prominent proposal for a global carbon budget is that we should emit no more than 1 trillion tons of carbon in total (Allen et al. 2009; Shue 2011). The grounds are these: roughly speaking, any more than a trillion tons would result in more than a 50 percent chance of exceeding a 2°C rise in global temperatures, where two degrees has been internationally agreed as a key target for avoiding dangerous anthropogenic interference with the climate system (UNFCCC 2009, 2015). Weisbach invokes the trillion ton proposal at one point, and observes that a mid-range projection for climate sensitivity would result in needing to get to zero somewhere between 2050 and 2060.[10] We note that this is part of the reason for the political focus on net zero by 2050–2060 in the countries mentioned earlier.

Why then does Weisbach not simply embrace this limit and say "no more than 1 trillion tons" or "no later than 2050–2060" (rather than "zero soon")? Notably, he neither (first) endorses this target, despite its prominence in mainstream climate policy circles; nor (second) does he provide a strong argument from self-interest for doing so. The most we see, late in the chapter, is a generic claim, familiar from the wider literature, that the downsides of severe climate change are so bad that it is a "better bet" to err on the side of caution than to take major risks. In short, while Weisbach has ample opportunity to push for a more aggressive approach than "zero sometime in the next century or so" in the chapter, he does not take up this opportunity.

Unfortunately, we suspect that he has good reasons for holding back, since we doubt that the simple self-interest approach can deliver more aggressive targets (like zero in 2050 or 2025). The first, more shallow reason is that the rationale for specific "scientific" targets are typically not as clear-cut or persuasive as one would like. In particular, it is not the case that there is a simple, decisive, scientific argument for a highly specific target. Instead, there is much divergence and variation.

One issue concerns the choice of appropriate objective. For example, over the last decade or so, many, especially in developing countries, have pushed for 1.5°C rather than 2°C, and a recent IPCC report argues that accepting 2°C involves allowing considerably worse impacts (IPCC 2018). Other, implicitly lower, targets have also been defended (e.g., 350 ppm for the atmospheric concentration).[11]

A second issue is that empirical estimates of what is required to meet specific targets vary widely. For instance, a recent overview states that estimates for the remaining carbon budget for two degrees range "from less than 800 Gt to almost 2,000 Gt CO_2," for 2018 onward, and "from below zero to more than 1,000 Gt CO_2," for 1.5 degrees (Lahn 2020, based on Rogelj et al. 2019).[12] These ranges are very large. Indeed, they are so large that it is unclear how well they can guide policy.

A third issue is that, apart from the empirical uncertainty, how one specifies and defends the objectives matters. For example, many endorse the one trillion ton target based on the idea that it is estimated to provide a 50 percent chance of achieving 2°C. Yet this is ethically questionable. Most notably, it is unclear why it is ethically defensible for us to accept anything as high as a 50 percent chance of breaching the target, and so of failing to prevent dangerous climate change. Others, for example, have proposed pushing for a 66 percent chance; and some may want better odds, such as 80 percent, 90 percent, or even higher. More importantly, we should avoid appearing simply to pick percentages out of a hat. Instead, what is needed is a good *justification* for tolerating any particular risk of failure, and for that justification to be robust against alternative justifications.

These issues suggest that choices of specific targets will be difficult for any view. Yet Weisbach has particular problems here, since he is aiming to offer an argument that is robust to *any* plausible assumptions about climate change, including scientific and economic assumptions.

This brings us to the second, deeper, reason why Weisbach might refrain from endorsing a specific, mid-century or earlier target. If he were to make the attempt, he would need to find a strong argument *from simple self-interest* in order to defend his choices over other choices (e.g., on temperature target, sensitivity, probability). Yet it is difficult to see where this argument would come from (see next section). In our view, this is not a mere oversight that is easily remedied. Ethics-based approaches might insist that certain kinds of risk are unacceptable, in part because the impacts that would result are too severe or harmful to be tolerated, on grounds of justice, welfare, or any number of ethical values. Yet it is unclear how a simple self-interest approach would do so.[13]

These problems probably explain why Weisbach does not take up the task for 2°C. Instead, he does his sensitivity analysis on impacts of up to 4°C, which he says all regard as truly catastrophic, since (for example) they imply the collapse of global agriculture (Ibid, 178; New et al. 2011). This sharply limits the scope of his argument, and is part of what drives the 100-year horizon. Yet even here it is difficult to make a simple self-interest argument based on 4°C using only conservative assumptions, as Weisbach aims to do. Mainstream climate science does not project a 4°C rise until late in the century, when many people around now will be dead. How then is it in their (simple, narrow) self-interest to stop emitting?

Speed

This leads us to the question of the *speed* of reductions. As we have seen, Weisbach claims that "it is in our self-interest . . . *to reduce emissions, starting now, and to do so rapidly*" (Gardiner and Weisbach 2016, 149). He also speaks of "drastic" emissions reductions, and "an energy transition as soon as possible." Given this, one would think Weisbach would favor a very quick transition to renewable energy, over a couple of decades, as many argue for (e.g., IPCC 2018; Shrader-Frechette 2011). However, this is not the case. In fact, Weisbach is scathing about such approaches. He dismisses those who argue for decarbonization over a decade or two as engaging in "sheer fantasy" that "perpetuates the myth" that energy transitions can occur in a short time period (ibid, 184).

In contrast, Weisbach's actual position is that we should "start now and *go slowly*" (ibid, 186). By this, he means that we should replace "existing fossil fuel infrastructure with clean energy *gradually as the existing infrastructure wears out*" (ibid, 186). He insists that this will be a slow process. He maintains that replacing the global energy supply with clean energy involves such a massive change in infrastructure that "*even a 100 year horizon . . . is not that long*" (ibid, 183).

Again, we see a mismatch between the rhetoric and the actual proposal. Weisbach's position is not particularly "aggressive." It calls for *slow* rather than "rapid," and *gradual* rather than "drastic," change. It also does not insist on reductions "as soon as possible," at least in any normal sense of the phrase.

Why does Weisbach believe that we must go slowly? There seem to be two reasons. The first is historical precedent. He claims that the last two energy transitions, from traditional biomass to coal and steam, and then away from coal to oil, gas, and electricity, took 80 and 130 years, respectively (ibid, 183). Interestingly this conclusion is in tension with the assessment of the IPCC, who say that the needed transitions are "unprecedented in terms of scale, *but not necessarily in terms of speed*" (IPCC 2018; emphases added).

Weisbach's second reason to go slowly is national self-interest, understood through his feasibility constraint. First, Weisbach asserts that countries with *existing fossil fuel infrastructure* are effectively committed to using it for its normal working lifespan, which turns out to be between 50 and 100 years. He says:

> How do leaders of a country explain that they are shutting down a plant that was just built at a considerable cost, works perfectly well, and is providing inexpensive and reliable energy to people who need it? *They can't. Once it is built, it will be used.* (ibid, 188);

and

> *Installing new fossil fuel infrastructure effectively commits* a country to the emissions from that infrastructure for its lifetime, which can easily be *fifty or even one hundred years.* Increasing the size of the fossil fuel infrastructure anywhere makes it more difficult to reach reasonable climate goals. (ibid, 188)

Second, he adds that *further, new investment* in fossil-fuel infrastructure is also inevitable:

> *To be sure, there is going to be some new fossil fuel infrastructure*, particularly in developing nations. *If we*—people or nations concerned about climate change—*tell India, China, or other fast-developing nations to scrap plans for new fossil fuel energy which they need for their economies to grow, we will simply be ignored.* (ibid, 188)

Is Weisbach correct to think that international climate policy is so sharply constrained? If he is, it is alarming news. As Weisbach points out elsewhere, the infrastructure is already in place that will use up most of the carbon budget for even a 50-50 chance of avoiding breaching the 2°C target for dangerous climate change (ibid, 185). For better chances, or to aim for the more ambitious and appropriate goal of a 1.5°C rise, matters are even worse.

Still, we believe Weisbach is not correct. We would argue that, given the severity of the climate crisis, closing down some (perhaps all) fossil-fuel infrastructure before the end of its usual lifespan is something that not only can and should be considered but also ought to be done. We also believe that this idea is now mainstream (e.g., Tong et al. 2018). We will revisit the putative constraints against this proposal later. For now, let us conclude that Weisbach's actual policy recommendations are notably less ambitious than other, more familiar proposals for decarbonization. Indeed, we would say that his approach lacks appropriate urgency.

III. SELF-INTEREST

Before saying more about Weisbach's feasibility constraint, let us first set the stage by considering his appeals to self-interest more generally.[14] We will argue that ultimately these appeals do not succeed. This is partly because his uses of self-interest shift back-and-forth between restricted and more expansive conceptions, and also between individual and collective (especially national) self-interest. We begin with the restricted interpretation, showing how it does not secure even the modest policy objectives Weisbach proposes. We then turn to more expansive conceptions, and raise the concern that they fail to offer a substantive alternative to ethical approaches.

The Simple Self-Interest Argument

Weisbach frames the basic climate problem as follows:

> Climate change is often portrayed as a very long-term problem, but *our children and grandchildren will be alive near the end of the century and into the next*, the time when we will face 2°C, 3°C, or even 4°C temperature increases. *If we use middle-of-this-century targets, many of the current adults will be alive.* While climate change will continue to affect people centuries in the future, it is a *surprisingly near-term problem. Climate change is about people alive today, their children, and their grandchildren* as well as the distant future. (ibid, 179)[15]

Weisbach appears to be *assuming* that the self-interest of all those involved *converges* on the same policies. It is explicit in the earlier passage that climate change is a problem both for those in the next three generations ("ourselves, our children, and our grandchildren"), *and* for people in the more distant future. But what if their interests do not align? It is not clear that this is the case, and Weisbach provides no argument for it. In addition, there are reasons to think that the convergence claim is questionable.

First, the most obvious worry is that the interests of *more distant generations* might diverge from those of the three currently alive. Suppose, for example, that those alive now engage in risky geoengineering that promises benefits for the next 100–200 years by holding off many bad effects of climate change but then makes matters much worse.[16] Such a policy would be in the interests of the next three generations (if only by hypothesis), but not the further future. We may think of it as the equivalent of an intergenerational time bomb. As such, it would be seriously unjust.[17]

Second, there is a less obvious, but perhaps deeper convergence problem (at least for the positive argument), recall that the positive argument appeals to a three-generation model: "people alive today, their children, and their grandchildren."[18] So, Weisbach needs a convergence of self-interest across the three generations at the heart of his position. But he does not argue for this. This is a major gap.

What kind of gap it is depends on how we understand the three-generation model. Let us consider two possible interpretations, both of which are of interest when understanding the self-interest approach, independently of what Weisbach himself thinks.

The first interpretation invokes what we shall call the *atomistic individualist model*. The idea is that each member of the three generations has *their own, independent argument* for robust climate action based on their own self-interest. In other words, the self-interest of all current adults demands climate action; so, independently and in its own right, does the self-interest of all of our children (many of whom are alive now); and so also (again independently and in its own right) does the self-interest of all of our grandchildren (some of whom are alive now). Moreover, and crucially, they all support more or less *the same kind* of climate action. Thus, on the atomistic individualist model, the self-interest argument needs to show:

A1. Robust climate policy X is in the self-interest of all current *adults*.
A2. Robust climate policy Y is in the self-interest of all *children* of current adults.
A3. Robust climate policy Z is in the self-interest of all *grandchildren* of current adults.
A4. Robust climate policies X, Y, and Z are *identical* (or close enough to make no practical difference).

Unfortunately, as far as we can tell, Weisbach provides none of these arguments. If he embraces the atomistic model, this is a major omission. It means that the heart of the argument is missing. We believe that this is a common problem across many self-interest accounts.

One reason the gap is important is that there is some presumption against a successful defense of the critical claims, (A1)–(A4). The first problem is the drift from individuals to single generations and then to multiple generations. Most generally, notice that on the surface the claims look very demanding. They cover *all* individuals across multiple generations and seek to maintain that all have exactly the same interests. Proving this *unanimity* claim is a very high bar to meet. For one thing, there seem to be counter-examples. Would the economic realist seriously argue that robust climate action is in the simple self-interest of a fossil fuel mogul in his seventies? We suspect not (if so, how?). But if it is not all, then who is covered?

Alternatively, perhaps what is intended is only a set of *majority* claims: that *most* adults, *most* children, and *most* grandchildren would benefit from robust climate action. This is more plausible; but then the realist needs to defend these weaker claims. They also need to explain why they are enough for the self-interest argument to deliver robust policy. For instance, there are many policies that might benefit most people that do not get enacted. Perhaps most people would benefit from a radical redistribution of global resources. But that does not make such a shift feasible or likely.

The second problem concerns difficulties in defending (A1), the claim that robust climate policy X is in the simple self-interest of current adults. Why assume this? The only metric that Weisbach mentions *is being "alive" when temperature targets are breached*. But there are numerous problems with this metric.

First, many adults alive now (in 2020) are old enough that they will not expect to face severe or catastrophic climate impacts themselves. Those currently over 80 or 90 may not expect to live much longer regardless; those over 70 may expect only to live another decade or two. In the US, for example, life expectancy is 78.9 years; in India it is around a decade less. In addition, many current adults will not expect to be alive beyond 2050 or so. Consider, for example, those over 50 in 2020. They will be over 80 in 2050, over 100 in 2070, and over 130 in 2100. So, if the truly severe or catastrophic impacts come in the second-half of the century, then it is not clear that it is in the simple self-interest of those already over 50 to avoid them. What then is the *uncontroversial* self-interest argument for the older generation supporting robust climate action? Arguably, they will not be around to reap most of the benefits. (There are, of course, ethical arguments that address these hurdles. We embrace them; but Weisbach and other economic realists want to avoid them.)

Second, even if many current adults will be around to experience severe or catastrophic impacts, this does not deliver the conclusion the self-interest argument needs. Weisbach appears to assume that if X is alive when the relevant threshold (e.g., 2°C, 3°C, or 4°C) is breached, then it is in X's *overall interests* to prevent that threshold being breached. Yet one's future life prospects do not necessarily define or dominate self-interest, nor do one's prospects in one's later years. For example, perhaps many affluent Westerners perceive their relatively high consumption lifestyles as more important to their self-interest than what happens toward the end of their lives. They prefer another decade or two of luxuriating in the high emissions world they are accustomed to, even given the threat of bad outcomes toward the end. Weisbach provides no argument as to why they should think differently, so that robust climate action is in their interests after all, relative to the alternatives.[19] (Again, we think that ethical arguments are needed.)

Third, the metric of being alive appears to make a dubious assumption: that using up a carbon budget and experiencing the effects of doing so occur at more or less the same time. Yet, climate change involves significant time-lags.

So, for example, as of 2020 we see a 1.2°C rise in the global average temperature; but conventional estimates suggest that we may already be effectively committed (for physical and/or social reasons) to as much as 0.5°C more.[20] Thus, it is possible that we have effectively breached the 1.5°C threshold already; nevertheless, it may still take a few decades for the temperature rise to manifest itself on the ground, and longer for the full effects to be realized (e.g., perhaps the melting of the Greenland ice sheet).

Similarly, although the carbon budget for 2°C may be exhausted by 2050, it will take longer for the full effects actually to be felt. Thus, using up the budget in 2050 does not mean that those alive in 2050 will then experience either a 2°C temperature rise or its full consequences *in 2050*. Both are likely to be deferred. Again, this changes the incentives, and in ways that threaten to undermine the simple self-interest claim. Put simply, there is an impact gap: if many of those alive when the carbon budget for 2°C is effectively breached *will not actually experience* the full effects, then one cannot *assume* that they have a (simple) self-interested reason to avoid the rise based on the *eventual* impacts; one can invoke only those impacts that will *actually accrue to them*. Again, their reasons to avoid the eventual effects will have to be about concern for others, and so are probably ethical in nature.

We conclude that Weisbach has not established that robust climate action is in the simple self-interest of all current adults. This is particularly true of older people in affluent countries. Since we might expect this group to hold disproportionate political power, they may be key to a self-interest-based solution. So, a gap there is very important (Gardiner 2004, 2011).

Unfortunately, this concern reverberates through Weisbach's whole argument. Worryingly, if the self-interest argument does not succeed with the older generation of the affluent, then Weisbach's positive argument is at risk of unraveling. This is particularly concerning when we turn to assumption

(4), that the robust climate policies allegedly favored by each generation X, Y, and Z are identical (or close enough to make no practical difference). For instance, in the extreme case, if the older generation does not favor truly robust climate action at all, then (4) is false and the simple self-interest argument falls apart.

More mildly, notice that even if Weisbach can make a positive case that *some* climate policy more aggressive than the status quo is in the interests of the older generation now, this does not guarantee that this policy will be strong enough to satisfy the next two generations. Consider a simplistic example just to make the basic point. Suppose we are talking about three generations of Northern Californians. All are affected by climate change. The grandparents (in their seventies) are subject to immediate risks of fire and smoke during fire season; the children (in their mid-forties) to these, their accumulated effects over decades, and to longer range effects of drought on agriculture; the grandchildren (in their teens) to all this and also to worries that the whole region may become unrecognizable over their lifetimes, with large areas that are inhospitable to contemporary ways of life that they cherish.

Suppose that, in response, each generation thinks solely in terms of their own, simple self-interest. First, the grandparents may then favor greater investment in fire prevention, or a move to an area with less flammable forest by the coast. They may remain unmoved by the need for mitigation; for them, by the time it will make a considerable difference they will likely be dead and gone. Second, their children, again thinking solely in terms of their own, simple self-interest, may favor more extensive mitigation, but only that designed to hold down temperatures over the next forty years or so. Moreover, since it is unclear how much can be achieved on that front given the time-lags involved, perhaps they think it is much better for them to move out of California to a state that is cooler or less susceptible to extreme impacts in the next few decades. Third, in contrast, the grandchildren will likely favor very deep emissions cuts, starting immediately and proceeding rapidly. After all, they will face many of the accumulated impacts of whatever emissions occur, and over many decades. They also have few alternatives. If climate change gets out of control and jumps to 4–6°C late in the century, they cannot plan simply to flee to another state. Given the physical impacts and the social upheaval caused by catastrophic climate change, there may simply be no place left to go (or no place that is at all "safe").

Familial Self-Interest?

Let us turn then to a second, more promising interpretation of the three-generation model, which we call the *familial inclusion model*. When Weisbach speaks of the self-interest of "people alive today, their children, and their grandchildren," perhaps he means to argue that these people understand their own interests in ways that mean that those interests are not independent of one another. The thought is that current adults conceive of their own interests as *including* those of their children and grandchildren (and perhaps that the later generations include their parents and grandparents in their conception of their own self-interest). Thus, the possibility of generational conflict is at least muted and perhaps removed. This is no longer "simple self-interest" but it retains something of the air of a self-interest account.

The familial inclusion model (of the three-generation approach) is more promising in part because it does seem difficult to argue that risking severe or catastrophic climate change is in the interests of current children and grandchildren. The threat is simply too large. At the high end, scientists see global temperature rises of 4–5°C or beyond as genuinely catastrophic, involving the large-scale collapse of agricultural systems and much else (e.g., New et al. 2011). Even the impacts of 2°C and 1.5°C are likely to be serious and widespread, which is why they are internationally accepted as constituting dangerous levels of climate change (e.g., IPCC 2018). Remember the impacts that we are already feeling, in terms of wildfires, floods, and hurricanes, even in the early periods of a 1.2°C rise. So, if the three generations really do care about one another and define their own interests collectively, then there may be a reasonable case for a "self-interest" argument on this basis.

Still, even on the inclusive model important ambiguities remain. For one thing, as stated, this expanded sense of self-interest includes merely *one's own* children and grandchildren, so that questions about convergence arise again. For example, why assume that a grandparent thinking of her children and grandchildren in a wealthy suburb in the US will desire the very same climate policy as a grandparent thinking of his children and grandchildren in the Brazilian Amazon? Perhaps both want action, and strong action at that, but they still disagree on the kind of action required. Perhaps one wants 2°C, zero by 2050, a strong military, and preparations for geoengineering; but the other wants 1.5°C, zero by 2030, and a return to a life living closer to the forest.

Given such issues, one might try to argue for stronger models. One possibility would be to extend the inclusion model beyond three generations, through a chain connection. So, for example, some might argue that, since our grandchildren will someday have their own children whose interests they adopt as their own, we, by extension, ought to adopt the projected interests of our grandchildren's children, and so on, *indefinitely into the future*. This is a new move, and more demanding. It involves defending what we might call an *indefinite chain inclusion model*. Another possibility is to argue for a model where each of the generations cares about the others *in their entirety*. This is a more radical move, and even more demanding. It involves defending a *national* or *global inclusion model*.[21]

To be clear, we are not necessarily against radical inclusion arguments. There are many possibilities, some of which are explored in the wider literature in intergenerational ethics.[22] Our main point here is merely that the discussion has now moved far beyond anything we'd be inclined to call *simple* or *narrow* self-interest, and these positions require robust defenses of their own.

Our own view is that self-interest is best understood as a substantive, complex and ethically loaded concept (Gardiner and Weisbach 2016, 64–65). This is especially true of notions such as national self-interest, and the self-interest of the current generation of humanity. Most notably, such concepts aim to integrate the concerns of millions, sometimes billions, of people, often over long periods of time. In doing so, they raise deep ethical questions, especially about justice. Take, for example, the genuine, substantive interests of the United States, understood as an entity spanning many generations stretching indefinitely into the future. Someone truly committed to the interests of this entity—to the American people thus understood—would, we think, have to favor strong climate action (Gardiner 2017b). But the commitment itself would involve strong ethical components, and so put significant pressure on Weisbach's claim to have provided a basis for climate policy that is independent of ethics.

It would be interesting to explore these questions further. That is a debate we would welcome. Alas, as we shall now see, Weisbach's actual argument appears to foreclose that debate, since its crucial component is a sharp feasibility constraint based on a much narrower account of *perceived* self-interest that threatens to overwhelm more robust conceptions.

Short-Term, Economistic Self-Interest

Unfortunately, whatever is going on with the three-generation model appears not really to matter much, at least for Weisbach, for he clearly believes that it is severely constrained in practice. Recall his argument for "going slowly," merely replacing existing infrastructure as it wears out, and accepting that some new fossil-fuel infrastructure "needed" for economic growth will still be built. In his view, constraints of political feasibility rule out any more aggressive approaches. He insists that once it is built, it will be used; that much fossil-fuel infrastructure has a lifetime of fifty to one hundred years; and that asking nations not to fuel economic growth in dangerous ways is unrealistic. More ambitious approaches, he says, merely indulge in "fantasy." In particular, the salient feature of "gradual replacement" is that "we *avoid the costs of scrapping perfectly good* power plants, refineries, tankers, and other parts of the fossil fuel infrastructure (ibid, 188)" and that if we ask "fast-developing nations to scrap plans for new fossil fuel energy which they need for their economies to grow, we will simply be ignored" (ibid, 188). Again, the key ideas rest on self-interest, and especially national self-interest. But this time it is *perceived* self-interest *defined in a narrow, short-term and economic way.*[23] Whatever sway any expansive, long-term, and three-generational conception of self-interest may have in Weisbach's argument, it appears to be swept aside by this rather dramatic constraint.

Regrettably, it is unclear how accepting the *short-term, economistic self-interest constraint* can lead to aggressive climate policies "far more ambitious than those currently on the table." Notably, Weisbach does not tell us how this tension can be overcome. The most we get is a vague gesture at the idea that if we start slowly, by replacing some existing infrastructure as it wears out, there will be technological advances "in the field" that will make investments cheaper later on (ibid, 188). To us, this seems a very complacent approach to climate policy, and a very risky bet for protecting against dangerous climate change. Surely, the magnitude of the climate problem justifies a much more aggressive approach.

One sign of the problem may be that, soon after introducing the idea that emissions must go to "zero soon," Weisbach himself suggests that even the more generous 2°C target for avoiding dangerous climate change "may be unrealistic" and "perhaps a more realistic target is 2.5°C or 3°C" (ibid, 177). This is not what one would expect if the positive argument were really driven by a robust commitment to the interests of "the children and grandchildren," but it becomes more plausible if, instead, the short-term, economistic constraint dominates.

IV. METHODOLOGY

How does Weisbach arrive at such a weak and disappointing climate policy? Our diagnosis is that the main drivers are a set of bold and contentious claims that emerge from a dubious methodology.

Contentious Claims

Let us begin with the claims. The first contentious claim is that "climate change is an energy problem" (179) involving "a massive change in infrastructure" (182). Weisbach asserts that "the central dilemma of climate change is straightforward: emissions have to go to zero while energy use has to remain high" (ibid, 182).

This framing of the climate problem is contentious because it *assumes* that solutions cannot involve reductions in global energy use. Thus, approaches that aim to reduce energy consumption (like those that retire infrastructure early) are automatically off the table. It is unclear why we should accept such a constraint. Policies that make energy use much more efficient, or replace energy-intensive activities with low energy activities, or that simply involve cut backs in energy use could make substantial contributions to climate action. A whole range of possibilities falls under these headings: from driving less and taking public transport more through to adopting simpler lifestyles with less emphasis on consumption. Why does Weisbach rule these strategies out?

The answer comes in terms of a second dubious assumption. Weisbach insists on what he calls "the iron law of wealth: increased wealth means increased energy use" (ibid, 181). Thus, his view is that high energy use is in everyone's self-interest, as it means increased wealth. This claim is contentious as well. At a general level, Weisbach is assuming that wealth must be our overriding objective and that it is appropriate to measure prosperity in terms of GDP per capita. Against this, we would insist that there are other values (including, for example, freedom, rights, community) that are also important to self-interest and of course to ethics more generally. We would also dispute that wealth and energy are central even when talking about well-being alone. For example, capability approaches seem to us to do better in capturing what really matters; and they are not so strongly indexed to energy use.[24]

Trends and Constraints

Still, rather than delving into these (common) disputes in detail, let us turn to a more general issue with Weisbach's methodology. Behind many of his most contentious claims lies a similar pattern of inference. Weisbach provides an (alleged) empirical trend or regularity, and then infers a policy constraint based on that trend. Three cases stand out. The first inference involves self-interest. The (alleged) trend is:

E1: Nations tend not to do what goes against their (national) self-interest.

The inferred policy constraint is:

C1: Nations will not do what goes against their (national) self-interest.

The second inference involves economic wealth. The (alleged) trend is:

E2: Reductions in wealth go against national self-interest.

The inferred constraint is:

C2: Policies and solutions cannot reduce national wealth.

The third trend is:

E3: High energy consumption is correlated with national wealth.

From this, the following policy constraint is inferred:

C3: Energy consumption must remain high.

In summary, in each case, Weisbach uses *trends* to justify *constraints*.

We now explore three concerns with this methodology. The first is that such trends need to be established empirically, and their limits explored. This is not a task that Weisbach and his colleagues typically take on; moreover, when they do, their defense is weaker than they admit. Either way, it would be rash to infer strong policy constraints on the future based on them, especially given what is at stake. The second concern is that even if such trends were empirically established, past trends are not necessarily a reliable guide to the future. The third concern is that this is especially so when the regularities being invoked are very close to the heart of the problem being confronted.

Empirical Support

Let us begin by assessing the empirical support Weisbach offers for his trends. Some regularities, such as those expressed by categorical laws, may form a constraint on action, e.g., the law of gravity. Weisbach's use of phrases such as "iron-law of wealth" (ibid, 181) suggests that he thinks empirical regularities E1–E3 may have such a law-like standing.[25] But the inference to a law-like regularity requires meeting a high bar of empirical support: not only must a correlation between variables be strong, it must rule out confounding factors and attempt to establish counter-factual variability through tests under a wide range of circumstances and conditions. Sometimes, Weisbach makes a law-like claim, but without the appropriate degree of empirical support.

Take, for instance, the inference to C3, the constraint that energy consumption must remain high. This relies on a correlation between national wealth and energy use which Weisbach calls the "iron-law of wealth" (ibid 181). Yet in reality, the empirical data he provides are a loose fit on a logarithmic scale (see graph 3.1).

Falling on Your Own Feasibility Sword?

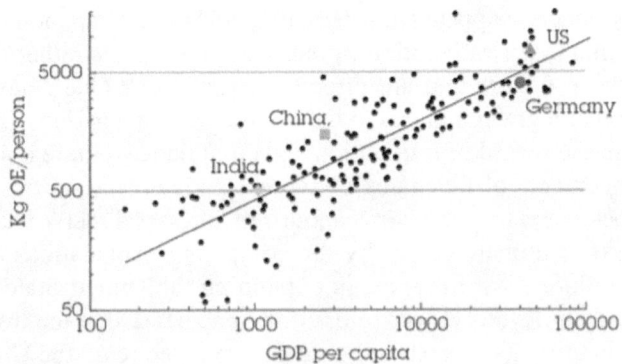

Graph 3.1 The relationship between Kg OE/person and per capita GDP.

The logarithmic scale distorts the (already substantial) amount of variability at the high end of consumption and GDP making it seem artificially tight. For instance, Germany and the United States' GDP per capita are comparable ($50,800 to $59,800) while Germany's energy consumption per capita is roughly half of the United States' (44,900 KWh v. 81,200 KWh).[26] Moreover, (and second) there is no discussion of, or attempt to defend, the independence of the data points. For instance, it seems just as likely that the correlation is an artifact of path-dependence or following the leader, rather than the result of multiple independent trials.[27] In addition, (third) even accepting the correlation at face-value, there is no serious attempt to rule out alternative explanations for it.[28] Finally, (fourth) these points bring up a more general issue with world-historical trends, which is that we have to base our analysis on one and only one trial. This severely constrains our ability to establish counter-factual variability.

The overreach in inferring an "iron law" from a suggestive scatter plot is especially surprising given the measured tone with which Weisbach elsewhere discusses the complex scientific phenomena behind climate change: "The amount of warming we will get is uncertain, and the science enormously complex. Notwithstanding years of work, we cannot pin it down" (ibid, 174); "The conclusions above were stated without repeated hedges about the vast uncertainty surrounding the effects of climate change. Climate science and climate change policy, however, are plagued by uncertainty" (ibid, 191). Apparently, the atmosphere is a complex system, while human societies can be understood and predicted with relative ease by looking at a scatter plot. We are skeptical.

Regarding other claims, it is unclear that Weisbach and company have even taken on the task of establishing the relevant trends empirically. Consider E1, the claim that nations tend not to do what goes against their (national) self-interest.[29] To the extent that this argument is backed by empirical evidence, it appears to offer a diagnosis or explanation of past failures of compliance or ratification in international treaties rather than noting a trend between factors that are directly observable.[30] One cannot observe directly whether a given treaty is in the self-interest of a nation. This must be inferred from the record of ratification and compliance, public opinion polls, the pronouncements of delegates or representatives, and so on. That such evidence does not establish claims about self-interest directly is also shown in the context of climate policy by the competing explanations offered for past policy failures. For instance, in explaining the United States' decision to withdraw from Kyoto we are told, on the one hand, that the treaty wasn't sufficiently in the self-interest of the parties involved (e.g., the United States anticipated unacceptable economic losses to ensue from participation). On the other hand, we hear that Kyoto failed because it didn't sufficiently constrain the self-interest of all participating parties (e.g., China, Brazil, India). Indeed, on accounts offered by ethicists, we should expect noncompliance when self-interest is insufficiently constrained. It is only the fair and equitable forgoing of each party's self-interest that can convince all countries to forgo their own narrow self-interest. In sum, E1 may attempt to provide an explanation for an empirical trend (to the extent that a handful of treaties and summits can establish these), but it can't claim to be the only or even the best explanation for these trends.

Projecting from the Past

We turn now to the second concern: the future reliability of past trends. Even if there is strong empirical support for the trends E1–E3 (a big if, we suspect), it would be rash to infer strong policy constraints on the future based on past trends, especially given what is at stake. Past trends are not necessarily a reliable guide to the future.

For one thing many philosophers of social science think that while we may be able to discover timeless regularities for "natural kinds," the kinds of objects studied by the natural sciences, the social sciences study interactive or "discursive kinds," namely human beings (e.g., Haslanger 1995; Hacking 1999). Discursive kinds are distinguished by their ability to interact with their own classifications, and even change those classifications over time. Hacking uses the example of the category: woman refugee (Hacking 1999). A woman refugee may be cognizant of her categorization as a woman refugee and act in ways that respond to that categorization and information about the category to change her own status (let's say to a citizen) and perhaps in the long run change the category and its implications more broadly (say, through a restructuring of immigration law). Applied to the cases at hand, perhaps noting that many of our past decisions have been made on the basis of considerations of narrow economic self-interest will, when brought to light, elicit a reappraisal of our priorities and reorientation of our practice toward their expression. We need not assume we are fixed in our own habits and thinking.

A more quotidian, but related, point is that trends in our behavior and their effects often depend upon some factor over which we do have control. Indeed, the hope that science can be used in a *critical* or *problem-solving* capacity must depend upon something like this assumption. For example, medical science may diagnose a regularity that indicates poor health, say, persistent high blood pressure. Luckily, we think that trend can be changed and in large part through varying factors that the patient is in control of, say, their diet and exercise habits. Theorists who appeal to feasibility constraints like Weisbach need to tell us why past regularities in energy consumption and international law, which have brought us to this problematic juncture, aren't like this. Indeed, if hasty inferences from loose past trends to inviolable future constraints are fair game, they must explain how their position allows any room for maneuvering our way out of the problem.

Entrenching the Problem

This brings us to the third issue: that the regularities turned constraints lie very close to the heart of the problem now being confronted.

Consider reasoning like the following invoked against feminist arguments for greater political and economic participation:

1. The US has never elected a female president. (trend)

Therefore,

2. Political parties should not put forward a female nominee. (constraint)

Or:

1. Men have always dominated the executive class. (trend)

Therefore,

2. Any feasible feminist policy must respect the constraint that men must continue to dominate the executive class.

In such cases, it is very troubling to see the very regularities that are part of the problem being invoked as constraints on the solution. For one thing, it seems to beg an important question. To assert that X is a constraint, is to claim that it is an ineliminable feature of the situation, at least for current purposes. But if the point of the whole enterprise is to change the situation, reasserting the regularity risks counting as obstructionism, or as an implicit demand that the status quo, or central components of it, remain in place.

This point is relevant to the climate context. Weisbach's claim is that climate action is only feasible if it is in accordance with national self-interest, understood in narrowly economic terms, and as correlated with high energy use. Yet, this comes close to insisting that factors that many maintain are at the heart of the problem—such as high consumption lifestyles, narrowly economic thinking, conventional state sovereignty (including institutional short-termism)—must remain unchanged. As a result, the apparent "constraints" risk *naming the problem*, rather than contributing to solutions. At the very least, they substantially reduce the available options or "tools in the toolkit." For example, notice how they rule out strategies that reduce energy consumption—say by substituting energy-rich consumption activities with more free time, improving national parks, and so on—right from the start.

V. CONCLUSIONS

We have argued that Weisbach's claim to offer a robust climate policy based on self-interest alone comes up disappointingly short.

One problem concerns ambition. The framing is of "aggressive" policies, "far more ambitious than those currently on the table," where this requires "rapid" and "drastic" emissions reduction, "an energy transition as soon as possible" and "zero or near-zero" emissions "in the near future." However, there is a serious mismatch between the rhetoric and the actual argument. The real plan is weak, tentative, and notably less ambitious than leading alternatives: we must start now, but go slowly; we are to think in terms of a hundred year timeframe; we can retire fossil-fuel infrastructure *only* when it no longer provides short-term economic benefits; we cannot prevent new infrastructure from being built that fuels growth.

We have offered a complex explanation of the roots of these disappointing results. One concern is that Weisbach falls on his own sword. His positive argument is held hostage to the same kinds of worries about political feasibility that he deploys against ethics- and justice-based approaches. Indeed, Weisbach insists on *such a tight set* of feasibility constraints that his argument calls into question the *very possibility* of appropriately ambitious climate action.

Another concern is that Weisbach oscillates between rival accounts of self-interest. He often emphasizes an expansive (we suspect ultimately ethical) conception of self-interest covering three generations; however, the notion of self-interest that actually drives his approach is very narrow, economic, and seemingly short-termist.

A third concern is that Weisbach uses a dubious methodology that is quick to infer hard constraints from weakly defended past regularities, and so lock us into the narrow bounds of the status quo. In doing so, economic realism threatens to hide indefensible behavior behind the pseudo-scientific language of "policy constraints."

We conclude that the simple self-interest argument is not only disappointing but also potentially dangerous. Like previous approaches to climate policy, it risks serving as a "dangerous illusion" that provides cover for an inadequate strategy that encourages further procrastination and delay.[31] This, of course, is far from Weisbach's intention. Still, given that he is a leading proponent of the self-interest view, the problems raise a challenge to the wider approach. For one thing, there is a burden on defenders of self-interested climate policy to provide a more successful positive argument for their position. For another thing, if (as we claim) the most important accounts of self-interest ultimately rely heavily on ethical concepts and ideals, climate advocates should be wary of being drawn into resistance to ethics and justice. If, as we suspect, there is no genuine alternative, attempts to marginalize ethics will be counterproductive for all concerned about protecting the future against severe climate change.

NOTES

1. Such views are not uncommon in policy circles, and also have some presence in the academic literature. In this chapter, we focus primarily on David Weisbach's position, as presented in Gardiner and Weisbach (2016). Other key work in the academic literature includes Posner and Weisbach (2010, 2013) and Posner and Sunstein (2008). While having a very different, and explicitly ethical approach, John Broome makes a proposal for "efficiency without sacrifice" which raises some overlapping issues (see Broome 2012 and, in response, Gardiner 2017a).

2. The precise nature of the resistance to ethics is complex and varies across proponents. For instance, in some places Weisbach says that he embraces a cosmopolitan approach to distributive justice (ibid, 154), and does not object to ethics playing a part in showing why climate policy is "desirable" (ibid, 152). Nevertheless, much of his work (alone and with Eric Posner) is devoted to arguing that justice has no role within climate policy (e.g., ibid, 151–152, 196; Posner and Weisbach 2013, 357), and his simple self-interest argument appears to foreswear any appeal to such values. For a wider discussion of the various—in our view, conflicting—strands of economic realism, see Gardiner in Gardiner and Weisbach (2016, chapter 3). For different concerns than those we raise in this chapter, see: Baer (2013); Jamieson (2013); Gardiner (ibid., chapters 3, 4, and 9).

3. All emphases are ours.

4. We also maintain that Weisbach fails to answer key questions about the viability of his approach more generally. Three failures are especially important. First, he does not explain how the self-interest approach is supposed to work internationally, particularly given his claim that major emitters must not be made worse off. Second, he does not address the criticism that his approach amounts to climate extortion. Finally, he avoids the intergenerational problem, and in particular the central challenge of the tyranny of the contemporary. See Gardiner and Weisbach (2016); Gardiner (2017a); Baer (2013); Jamieson (2013).

5. Gardiner and Weisbach (2016, 170–172).

6. Our definition differs slightly from the norm in that it makes clear that the carbon budget takes a position on acceptable risk. In our view, this is a key component.

7. For example, we do not see Weisbach showing how simple self-interest justifies "zero eventually" with low-end projections of carbon sensitivity and climate harms, or the argument from simple self-interest for net zero for those of advancing years (say, over 70, 60, or even 50).

8. E.g., IPCC (2018), Headline Statements from the Summary for Policymakers: "In model pathways with no or limited overshoot of 1.5°C, global net anthropogenic CO_2 emissions decline by about 45% from 2010 levels by 2030 (40–60% interquartile range), reaching net zero around 2050 (2045–2055 interquartile range). For limiting global warming to below 2°C CO_2 emissions are projected to decline by about 25% by 2030 in most pathways (10–30% interquartile range) and reach net zero around 2070 (2065–2080 interquartile range)," 2.

9. See also Gardiner (forthcoming).

10. Gardiner and Weisbach (2016, 179).

11. E.g., Hansen et al. (2008).

12. Rogelj et al. (2019, 338), as cited by Lahn (2020).

13. Weisbach does invoke the idea of "unacceptable harms" in some places (e.g., ibid 146, 149). However, for his positive argument to work these harms must be understood purely in terms of narrow or simple self-interest. So, for example, he cannot claim merely that temperature rises of more than two degrees would cause great harm or constitute grave injustices for large numbers of people (as an ethics-based view can). He must *also* claim that the infliction of these harms and injustices are bad *for the simple self-interest* of *all* those affected, by both the harms themselves *and* (crucially) the policies necessary to avoid them. He must show too that they are worse than the alternatives. This is a tall order. For instance, one reason to think the task is more difficult for a self-interest approach than one based on ethics is that we usually think it is permissible to take more risks with one's own interests than with the interests of others to whom one has moral and political responsibilities.

14. Weisbach does not offer definitions of self-interest or simple self-interest. He appears to regard definitions as unnecessary for his purposes. Presumably, this is because he believes that his argument is robust to a broad range of assumptions, including about self-interest itself.

15. One (minor) point is that we find this framing potentially misleading. As far as we are aware, no one disagrees with the claim that climate change is a problem that affects people alive today as well as many yet to be born. This is common ground between ethics-based and self-interest-based approaches. Instead, what is different about the self-interest approach is that it seeks to ground a robust climate policy *on the basis of self-interest alone*. Presumably, this shifts the discussion to people alive today. It is only their self-interest that is relevant.

16. Gardiner (2011).

17. As far as we can tell, Weisbach does not even mention the *possibility* of such divergence when making his argument in the book. This might be understandable for some economic realists, who may believe that only the self-interest of the

three generations matters. But we think taking this attitude is a mistake. An ethical approach would include concern for the future of our offspring and of humanity as such over a much longer timeframe. It would reject intergenerational time bombs. If a self-interest approach cannot deliver this, then it is deficient as a basis for climate policy.

We suspect Weisbach himself would agree, and accept this burden of proof. However, in the book, all we see is a footnote that states that "the argument in the text focuses on people alive today and their immediate descendants," and that he believes that "self-interested choices fully take future generations into account" (168, note 22). The footnote implies that his previous work on discounting demonstrates this. We do not believe that is the case, or that discounting is the core issue. Still, like Weisbach, we will pass up the opportunity to discuss discounting directly and instead simply refer to work elsewhere (e.g., Gardiner 2011). Our point here is simply that in presenting his core positive argument for the simple self-interest approach, Weisbach chooses to avoid an issue right at the heart of the matter (convergence with future generations) and does so by invoking a controversial position in an area that is itself deeply controversial (discounting). This avoidance does not bode well for the ambition of offering a simple and uncontroversial argument based on self-interest alone. (In contrast, we think that the intergenerational issue is a central one in climate ethics, since there is a threat of a tyranny of the contemporary (Gardiner 2011, 2019).)

18. For example: "even using narrow notions of self-interest, it is in our self-interest to wisely govern use of the atmosphere, to reduce emissions, starting now, and to do so rapidly. We need to do so to prevent very serious harms to ourselves, our children, and our grandchildren" (Gardiner and Weisbach 2016, 149).

19. Perhaps he means to claim that the impacts in later life (even very late life) are so very bad that they outweigh earlier gains in terms of simple self-interest alone. We have two responses. First, this claim is more plausible for high impacts coming soon (e.g., 4 degrees by 2050) than it is for more moderate impacts coming slowly (e.g., 2 degrees by 2100), but Weisbach wants to be robust to a wide range of scenarios. Second, Weisbach is talking about perceived self-interest, which is notoriously bad at tracking real self-interest, especially in matters concerning the future (e.g., consider smoking). Those arguing on the basis of ethics and justice can appeal to the grave wrongs high consumption and indifference impose on others, and also perhaps to the implausibility of the shallow conception of self-interest that is being assumed. But Weisbach has no such recourse. He asserts that his view is based on "simple" self-interest, presumably understood in uncontroversial ways.

20. We might be effectively physically committed due to (for example) past emissions plus a very high climate sensitivity; we might be effectively socially committed due to (for example) future emissions that we will no longer be able to avoid for political, economic, or ethical reasons.

21. Weisbach's approach sometimes seems to drift in this direction. For example, he says: "The outer bounds of reasonable climate policies act also as *constraints on ethics*. Policies that do not stay within the bounds *risk serious harms*. In particular, any policy that does not require emissions reductions by all nations in the not-too distant future will violate *basic imperatives to avoid serious harms* from climate change"

(Gardiner and Weisbach 2016, 197–198). How are we to understand these "harms" and the "basic imperatives" to avoid them? Weisbach's remarks suggest that he sees them as *prior to* and constraints on ethics. But we submit that they are most naturally understood in ethical terms, and that it is difficult to make sense of them without embracing ethics. Thus, if an expansive three-generation model includes concern for *harm* to (very many) others, in so doing it begins to blend with the supposedly unnecessary and independent ethical perspective. In any case, at this point perhaps the concept of self-interest has become largely superfluous. If economic realists are forced to appeal to notions of harm and broad intergenerational concern in the end anyway, we wonder what extra value talk of self-interest brings to the table.

22. E.g., Meyer and Gosseries (2009).

23. In earlier work, Posner and Weisbach characterize their feasibility constraint by saying: "We believe that states (and not just the US) define their self-interest in narrower terms, oriented mainly toward wealth and security" (Posner and Weisbach 2013, 357); that this "probably requires that all states … are *economically* better off" (Posner and Weisbach 2010, 179); and that "past climate policy does not depart from nations' perceived *short-term* self-interest" (Ibid, 72). They argue that the definitional constraint is powerful: for example, it "*rules out*" the United States coming to accept "a moral obligation to bear the bulk of the climate burden because the US is wealthier than most other states, and because it is responsible for a large portion of the greenhouse gas emissions in the atmosphere." They add that "our conception [of the constraint] is based on our reading of history … we do not expect Americans (or people in other countries) to define their national interest so capaciously [as to include such moral obligations] because they never have in the past." (Posner and Weisbach 2013, 357). We disagree with this assessment. Our point here, however, is much narrower. It is that the way Weisbach operationalizes the constraint in the simple self-interest argument threatens to undermine a truly ambitious approach of the kind needed.

24. For capabilities approaches, see Sen (1999) and Nussbaum (2011).

25. Gilabert and Lawford-Smith (2011) characterize nomological constraints as "hard" constraints, which rule actions out as opposed to soft constraints—things like motivational or cultural factors—which make actions comparatively more or less feasible, but don't rule them in or out categorically. The authors claim that soft constraints are "[t]he more salient and difficult constraints to consider in political discussions" (p. 813), though they do not argue for this claim. The following discussion will hopefully make clear that a justification is needed since there are feasibility theorists arguing that laws can do much of the work in ruling out various policy options. Along the way we will try to provide some of the missing argument for thinking that social or economic laws won't in fact do this work.

26. Ritchie (2014) and CIA World Factbook (2020).

27. Regarding path-dependency, for instance, new infrastructures are often constructed using parts or key features of old ones (thus retired railroad right of ways get recycled as motor-vehicle roads while keeping other features of, say, food distribution or transport networks fixed). This piecemeal way of development takes the current path of least resistance available to individual entrepreneurs or regional

authorities, which often depends upon the previous infrastructure, e.g., converting from rail to truck. But it does not signify the only possible transformation. With greater coordination, other features of a distribution system such as average shipping distance might change. Concerning the claim that the data may be explained by the phenomenon of following the leader there is also a robust record of international exchange of technologies, ideas, and personnel that goes into the spread of infrastructure globally. (For a history of the exchange of ideas and patents among Western European countries and the United States in electricity infrastructure, see Hughes 1983).

These considerations suggest the data are not best understood as the result of countries independently trying to maximize wealth by a variety of means. Rather the actual course of historical development was both highly dependent upon what came before and globally intertwined.

28. On both of these points, see Yergin (1991), on the history of oil exploration and development. One of the main drivers of the adoption of modern forms of fossil-fuel energy, especially oil, was the military advantage that it gave countries during World War I and World War II. Subsequent to the wars, much of this infrastructure transitioned to civilian, economic use. Thus, some measure of militarization (including alliances) might be a confounding factor.

29. In some places, E1 appears to be based on a conceptual connection between notions of voluntary participation and self-interest. We doubt the conceptual argument works. For instance, Posner and Weisbach (2010) cite Jonathan Wiener (1999). Wiener argues that climate change negotiations will take the form of an international treaty and that such treaties rely upon voluntary consent of the parties involved (based in turn on the assumption that the coercive mechanisms that might enforce compliance are relatively weak in the international context). Voluntary compliance is supposed only to be possible if the treaty makes each country no worse off (according to that country's conception of its self-interest). While this inference may work for a very broad sense of self-interest, it can't be used to infer the claim that nations act in their self-interest where this sense of self-interest excludes ethical considerations. Would nations think of themselves as worse-off because they agree not to commit genocide for purely ethical reasons (and not merely, say, to avoid risks to their own populations)? Does their participation on such grounds thereby become involuntary? We doubt it. In practice, conceptions of national self-interest are, in part, ethical. Who we (as a nation) are, ethically speaking, matters to us. And this opens up the possibility of voluntary participation on ethical grounds.

30. Another thing worth flagging here, is that the supposed regularity only applies within the context of international treaties. But an international treaty may not be the only available instrument. Given the scope and severity of the problem, it may be time to explore alternative legal tools such as setting up global institutions with greater authority (e.g., Gardiner 2014, 2019).

31. Gardiner (2004, 2011). As a result, it may encourage moral corruption, that is ways of thinking and talking about the climate problem that obscure many of its most important features, including threats of serious injustice.

REFERENCES

Allen, Myles, David Frame, Chris Huntingford, et al. 2009. "Warming Caused by Cumulative Carbon Emissions Towards the Trillionth Tonne." *Nature* 458(April): 1163–1166.

Baer, Paul. 2013. "Who Should Pay for Climate Change? 'Not Me.'" *Chicago Journal of International Law* (13)2: 507–525.

Broome, John. 2012. *Climate Matters*. New York: Norton.

CIA. 2020. *The World Factbook 2020*. Washington, DC: Central Intelligence Agency.

Gardiner, Stephen. Forthcoming. "Targets."

Gardiner, Stephen. 2004. "The Global Warming Tragedy and the Dangerous Illusion of the Kyoto Protocol." *Ethics and International Affairs* (18)1: 23–39.

Gardiner, Stephen. 2011. *A Perfect Moral Storm*. New York: Oxford University Press.

Gardiner, Stephen. 2014. "A Call for a Global Constitutional Convention Focused on Future Generations." *Ethics and International Affairs* (28)3: 299–315.

Gardiner, Stephen. 2017a. "The Threat of Intergenerational Extortion: On the Temptation to Become the Climate Mafia While Masquerading as an Intergenerational Robin Hood." *Canadian Journal of Philosophy* (47)2–3: 368–394.

Gardiner, Stephen. 2017b. "Trump and Climate Justice." *The Philosopher's Magazine*. https://www.philosophersmag.com/essays/168-trump-and-climate-justice.

Gardiner, Stephen. 2019. "Motivating a Global Constitutional Convention for Future Generations." *Environmental Ethics* 41: 199–220.

Gardiner, Stephen M., and David A. Weisbach. 2016. *Debating Climate Ethics*. New York: Oxford University Press.

Gilabert, Pablo, and Holly Lawford-Smith 2011. "Political Feasibility: A Conceptual Exploration." *Political Studies* 60: 809–825.

Hacking, Ian. 1999. *The Social Construction of What?* Cambridge: Harvard University Press.

Hansen, James, Makiko Sato, Pushker Kharecha, David Beerling, Robert Berner, Valerie Masson-Delmotte, Mark Pagani, Maureen Raymo, Dana L. Royer, and James C. Zachos. 2008. "Target Atmospheric CO_2: Where Should Humanity Aim?" *The Open Atmospheric Science Journal* 2: 217–231.

Haslanger, Sally. 1995. "Ontology and Social Construction." *Philosophical Topics* 23(2): 95–125.

Hughes, Thomas. 1983. *Networks of Power*. Baltimore: Johns Hopkins University Press.

IPCC (Intergovernmental Panel on Climate Change). 2018. *Global Warming of 1.5°C. An IPCC Special Report on the Impacts of Global Warming of 1.5°C Above Pre-Industrial Levels and Related Global Greenhouse Gas Emission Pathways, in the Context of Strengthening the Global Response to the Threat of Climate Change, Sustainable Development, and Efforts to Eradicate Poverty*, edited by Valérie Masson-Delmotte, Panmao Zhai, Hans-Otto Pörtner, Debra Roberts, Jim Skea, Priyadarshi R. Shukla, Anna Pirani, et al. In Press.

Jamieson, Dale. 2013. "Climate Change, Consequentialism, and the Road Ahead." *Chicago Journal of International Law* (13)2: 439–468.
Lahn, Bard. 2020. "A History of the Global Carbon Budget." *Wiley Interdisciplinary Reviews: Climate Change* 11.
Meyer, Lukas H., and Axel Gosseries, eds. 2009. *Intergenerational Justice*. Oxford: Oxford University Press.
New, Mark, Ruth I. Hart, Barbara Kimbell, Janet M. Allen, Rachel Elizabeth Jane Besser, Charlotte Boughton, Daniela Elleri, J. Fuchs, A. Ghatak, T. Randell, and A. Thankamony. 2011. "Four Degrees and Beyond: The Potential for a Global Temperature Increase of Four Degrees and Its Implications." *Philosophical Transactions of the Royal Society A: Mathematical, Physical and Engineering Sciences* (369)1934: 6–19.
Nussbaum, Martha. 2011. *Creating Capabilities*. Cambridge: Harvard University Press.
Posner, Eric A., and Cass R. Sunstein. 2008. "Climate Change Justice." 96 *Georgetown Law Journal*: 1565–1612.
Posner, Eric A., and David A. Weisbach. 2010. *Climate Change Justice*. New Jersey: Princeton University Press.
Posner, Eric A., and David A. Weisbach. 2013. "International Paretianism: A Defense." *Chicago Journal of International Law* (13)2: 347–358.
Ritchie, Hannah. 2014. "Energy." *OurWorldInData.org*. https://ourworldindata.org/energy.
Rogelj, J., P. M. Forster, E. Kriegler, C. J. Smith, and R. Séférian. 2019. "Estimating and Tracking the Remaining Carbon Budget for Stringent Climate Targets." *Nature* 571(7765): 335–342.
Sen, Amartya. 1999. *Development as Freedom*. New York: Anchor.
Shrader-Frechette, Kristin. 2011. *What Will Work*. New York: Oxford University Press.
Shue, Henry. 2011. "Human Rights, Climate Change and the Trillionth Ton." In *The Ethics of Global Climate Change*, edited by Denis Arnold, 292–314. Cambridge: Cambridge University Press.
Tong, Dan, Qiang Zhang, Yixuan Zheng, Ken Caldeira, Christine Shearer, Chaopeng Hong, Yue Qin, and Steven J. Davis. 2018. "Committed Emissions From Existing Energy Infrastructure Jeopardize 1.5°C Climate Target." *Nature* (572): 373–377.
UNFCCC (United Nations Framework Convention on Climate Change). 2009. "Report of the Conference of the Parties on its fifteenth session, held in Copenhagen from 7 to 19 December 2009." https://unfccc.int/resource/docs/2009/cop15/eng/11a01.pdf.
UNFCCC (United Nations Framework Convention on Climate Change). 2015. "Adoption of the Paris Agreement." Decision 1/CP.21. Document FCCC/CP/2015/10/Add.1. Paris.
Wiener, Jonathan. 1999. "Global Environmental Regulation: Instrument Choice in Legal Context." *Yale Law Journal* 108(4): 677–800.
Yergin, Daniel. 1991. *The Prize*. New York: Simon & Schuster.

Chapter 4

Climate Justice, Feasibility Constraints, and the Role of Political Philosophy

Brian Berkey

Philosophers doing normative work on urgent matters of great moral importance are often motivated to pursue that work at least in part because they believe that it has the potential to contribute to real world progress on the issues about which they are writing. Many outside of philosophy are inclined to think that this is the only type of reason that could justify engaging in such work. While philosophers tend not to have quite so narrow a view about the reasons that count in favor of pursuing philosophical projects, it is not surprising that many who work on issues of great moral importance (e.g., global poverty, racial justice, animal rights, climate change) have as one aim of their efforts making a positive difference with respect to the issues that are their focus.

The idea that normative philosophical work should be capable of positively impacting real-world decision-making regarding the issues addressed has contributed in recent years to increased skepticism of some traditional modes of normative theorizing.[1] In particular, certain approaches to thinking about justice have been challenged on the grounds that they tend to imply that policies and outcomes that are (virtually) certainly never to be enacted or achieved are required by justice. Skepticism of this general kind has led some philosophers to claim that considerations of feasibility ought to constrain theorizing about justice. Feasibility constraints imply that an argument to the effect that a policy or outcome is a requirement of justice should be rejected, even if it is otherwise normatively appealing, if enacting the policy or bringing about the outcome is, in the relevant sense, infeasible.

Feasibility constraints can be more or less constraining, depending on the conditions that are taken to determine the relevant sense of feasibility. On some accounts, the feasible set will be quite large, and the departures from traditional modes of normative theorizing required by the associated feasibility constraints will be correspondingly modest. On other accounts, however, the feasible set will be more limited, and the departures from traditional modes of theorizing required will be significantly greater.

In recent discussions of climate justice, some theorists have suggested that we should accept fairly substantial feasibility constraints on our theorizing (e.g., Posner and Weisbach 2010; Brandstedt and Bergman 2013; Gajevic Sayegh 2019). My central aim in this chapter is to argue that even if we accept that normative work on urgent issues such as climate change ought to be capable of contributing in a practical way to efforts to address those issues, there are strong reasons to reject these feasibility constraints. There are, I will claim, a number of valuable practical roles that philosophical theorizing that is not done within the limits of such constraints (henceforth "ambitious theorizing") can play, even in urgent circumstances like those that we currently face with respect to climate change.

I will proceed in the remainder of the chapter as follows. First, I will briefly highlight some central features of the current climate crisis. In light of these features, I will describe a plausible initial argument for radical requirements of climate justice. And I will note several policies that might be advocated as potential means of satisfying (or at least increasing the satisfaction of) those requirements. Next, I will note several grounds on which some may raise feasibility-based objections to the argument for radical requirements of climate justice, and I will argue that there are clear limits to what we can plausibly take these objections to support with respect to the content of climate justice. I will then describe in greater detail some of the central features of the debate about the place of feasibility constraints in political philosophy. I will provide grounds for thinking that some proponents of such constraints, and in particular some contributors to recent discussions of climate justice and policy, endorse the claim that ambitious theorizing has no valuable role to play in urgent circumstances. Finally, I will respond to this claim by describing the valuable roles that I believe ambitious philosophical work can play in the struggle against climate change.

My aims can plausibly be viewed as fairly modest. The success of my argument requires only that I identify sufficiently valuable roles that some ambitious theorizing about climate justice can play in the struggle to limit dangerous climate change.[2] Nonetheless, I believe that this is an important and worthwhile task to take on, in light of the increased skepticism about the value of such work that has been expressed in recent years, as well as the importance of doing everything that we can to advance the cause of climate justice.[3]

I. THE CLIMATE CRISIS

In the thirty years since the Intergovernmental Panel on Climate Change released its first comprehensive assessment report (IPCC 1990), shockingly little has been accomplished in terms lowering greenhouse gas (GHG) emissions trajectories.[4] It is widely agreed among scientific experts that we now face a situation in which large net emissions reductions are required in a relatively short period of time if we are to give ourselves a reasonable chance of limiting the global mean temperature increase to no more than the generally accepted target of 2°C above pre-industrial levels (Working Groups I, II, and III 2015). If we are to aim for the more ambitious target of limiting warming to no more than 1.5°C, which many have suggested we have strong reasons to do (IPCC 2018; Alliance of Small Island States 2019), then even more dramatic reductions are required in an even shorter period of time.[5]

If, on the other hand, we continue to avoid increased mitigation efforts, it is estimated that the global mean temperature increase will likely reach between 3.7°C and 4.8°C by 2100, although it could be as high as 7.8°C (Working Groups I, II, and III 2015, 20). These levels of warming would all have catastrophic consequences, with the expected consequences getting worse as the amount of warming increases. Among the familiar predicted and potential consequences of warming that reaches well above 2°C are widespread famine, severe and deadly heatwaves, the destruction of coastal lands due to sea-level rise, mass migration, and global resource wars (United States Department of Defense 2015; Wallace-Wells 2019).

II. AN INITIAL CASE FOR RADICAL REQUIREMENTS OF CLIMATE JUSTICE

It is clear that the consequences of continued inaction in response to the climate crises would be disastrous. Because of this, we must recognize and take seriously that radical changes are required in order to meet the targets endorsed by scientific experts. The domains in which such changes are required plausibly include energy and economic policy, land use, corporate decision-making, and behavior and individual lifestyle choices. Failure to make the necessary changes would almost certainly have disastrous consequences for future generations of the kinds noted earlier.

Given these facts, it would seem that even rather modest and widely accepted claims about the nature and grounds of justice between generations imply that radical changes are required of us as a matter of justice. If, in addition, we accept even rather modest and widely accepted claims about the justice-relevant considerations that bear on how the burdens of required mitigation efforts ought to be distributed, then we appear to be committed to accepting that justice requires the well-off to accept potentially very large sacrifices, whether directly or as a result of required policy changes (or, perhaps most plausibly, a combination thereof).[6] Consider, for example, the implications of the claim that mitigation costs should not be imposed on the global poor if this would undermine or delay their prospects for escaping poverty, at least so long as the relevant costs could instead be taken on by better off people (Moellendorf 2014, chap. 1).

A strong case, then, can be made for thinking that some very radical policy and associated behavioral changes are required, as a matter of justice, in wealthy and high-emitting countries like the United States. While I will not attempt to defend any view about exactly which policies we ought to endorse as required by justice or aim to promote in our political activities, it will be helpful to have some examples in mind for illustrative purposes. Consider, then, the following policies:

1. A carbon tax on businesses, set at the high end of current estimates of the social cost of carbon,[7] with the revenue generated used to (perhaps among other things) offset any increase in the cost of essential goods for the poor. (For more on this topic, see Meyer, K. in this volume.)
2. A ban on new exploration for fossil fuel deposits, in combination with strict and increasing restrictions over time on the extraction of known sources of fossil fuels.[8]
3. Massive public investment, funded by large tax increases on wealthy citizens and/or corporations,[9] in renewable energy research, development, and infrastructure, including aid to poor countries aimed at facilitating the adoption of renewable technologies rather than fossil fuels.
4. Official efforts to promote, and clear willingness to ratify, a global climate treaty that includes binding and enforceable commitments to reduce emissions by an amount that plausibly reflects each country's fair share of the burdens of meeting the required global emissions reduction targets, within the time frame necessary to do so.

It seems likely that if these policies, or any subset of them, were adopted and complied with, then emissions would be reduced, perhaps substantially. At the very least, this would be the case if the policies were combined with certain patterns of behavior among individuals and corporations within the policy constraints. Compliance with these or similar policies, in combination with other behavioral changes, would almost certainly contribute to advancing climate justice by making it more likely that warming is limited to less than 2°C (or 1.5°C), or at least minimizing any warming beyond those targets.

III. FEASIBILITY OBJECTIONS

Despite the normative appeal of the claim that a country like the United States is required, as a matter of justice, to adopt policies such as (1)–(4), or similarly ambitious policies, some would object that this cannot be the case because their adoption is, in some relevant sense, infeasible. They might, for example, insist that enacting these policies is not feasible due to the fact that many of those who would face costs if they were adopted possess disproportionate political influence, and will surely oppose them. Or, they might claim that the policies could not feasibly generate the emissions reduction and other climate justice-promoting effects that their proponents intend, due to the ways that individuals and corporations would in fact behave if they were enacted.[10] Some who hold that considerations of feasibility constrain what can be required as a matter of justice might even think that these responses constitute sufficient grounds for concluding that adopting the policies (or any subset of them) is not required by justice.

Accepting this, however, would commit us to accepting some rather troubling implications, and even most philosophical proponents of feasibility constraints on requirements of justice would not accept them.[11] This is because the mere unwillingness of some people to behave in ways that are required in order for a purported requirement of justice to be met cannot, in and of itself, provide grounds for concluding that it is not a requirement of justice after all (Räikkä 1998, 29; Gilabert and Lawford-Smith 2012, 813; Lawford-Smith 2013, 256; Roser 2016, 84; Gilabert 2017, 105). With respect to climate justice, accepting that policies like (1)–(4) are not required for feasibility-based reasons would also seem to commit us to thinking either that limiting warming to less than 2°C (let alone 1.5°C) cannot be required as a matter of justice, or that it is consistent with justice to, for example, impose significant costs of mitigation on the global poor of current and/or future generations in order to limit warming. After all, the very same motivations and behavioral dispositions that would be taken to explain why enacting the policies and/or using them to generate large emissions reductions is infeasible would also presumably make it infeasible to sufficiently reduce emissions by any other means that does not involve compensating well-off high emitters for reducing their emissions.[12]

I assume (but will not argue in this chapter) that any account of the fundamental requirements of climate justice that does not imply that we must (without avoidably imposing costs on the global poor) meet either the 2°C or 1.5°C target, or that one of these targets, in combination with the best scientific evidence, determines the extent of the emissions reductions that are required by justice,[13] is unacceptable.[14] However unlikely it is that we will in fact meet either the warming limitation targets or the emissions reduction targets, meeting them is not infeasible in any sense that could plausibly make it the case that we are not required as a matter of justice to do so.

Those who think that considerations of feasibility ought to inform philosophical theorizing about justice might accept this as a matter of principle.

They might, nonetheless, insist that, at least in our current urgent circumstances, there is no (sufficiently) valuable role for philosophical theorizing about climate justice that does not aim to recommend actions, policies, and goals that might actually be adopted or achieved. The urgency of making whatever progress we can on mitigation (see e.g., Moellendorf 2016, 105–107), in combination with the dismal record that we have amassed over the past three decades, they might argue, should lead us to think that much theoretical work on climate justice has little, if anything, of value to offer in the struggle to prevent, or at least limit, dangerous climate change. Claims of roughly this kind are often appealed to in the growing body of work criticizing "ideal theory" in political philosophy more generally, and also seem to underlie some recent work on climate justice and policy (e.g., Posner and Weisbach 2010; Brandstedt and Bergman 2013; Gajevic Sayegh 2019).

IV. FEASIBILITY CONSTRAINTS ON THE CONTENT OF JUSTICE AND FEASIBILITY CONSTRAINTS ON THEORIZING IN URGENT CIRCUMSTANCES

While the practical concerns that I have noted, and that are my central focus in this chapter, clearly lie behind some of the recent criticism of ambitious theorizing, much of the recent debate in political philosophy about whether we ought to accept feasibility constraints on the content of requirements of justice focuses on somewhat more abstract issues. In these discussions, the range of possibilities that are thought of as feasible is generally limited only by the principle that "ought implies can." One central point of contention in these debates is, for example, whether the concept of justice should be thought of as constrained by the "ought implies can" principle at all (Gheaus 2013; Wiens 2014), so that outcomes that, in the relevant sense, we could not bring about cannot be required as a matter of justice. Another is precisely how the "ought implies can" principle should be interpreted, assuming that it does constrain the content of requirements of justice (Estlund 2011b, 2016; Wiens 2016; Chahboun 2017).

There are, however, some who have suggested that there are reasons to endorse significantly stricter feasibility constraints on theorizing (Brennan and Pettit 2007; Brennan 2013; Hamlin 2017, 222–226).[15] Most commonly, the stricter constraints are thought to be grounded in the findings of social scientific research, and in particular economics.[16] Geoffrey Brennan and Philip Pettit claim, for example, that political philosophers ought to take what they call "compliance constraints" into account in their theorizing (2007, 260), in order to ensure that their recommendations are "incentive compatible" (ibid., 264). Compliance constraints limit how effectively various policies can be expected to advance the justice-based goals that might initially be thought to justify their adoption. A policy that faces serious compliance constraints might be such that, although it would be the best possible policy in conditions in which people fully comply with both its letter and broader expectations, it would do much less to advance justice in the actual world than alternatives, since compliance levels will in fact be far from optimal (ibid., 260). Failure to take compliance constraints, and the associated requirement of incentive compatibility, sufficiently seriously, Brennan and Pettit claim, "is capable in principle of undermining an entire normative enterprise" (ibid., 260).[17]

In the debate about climate justice, a number of theorists suggest that rather strict feasibility constraints should limit our theorizing. For example, Eric Brandstedt and Anna-Karin Bergman claim that one of the two basic desiderata for an account of climate rights is "its ability to generate political measures" (2013, 396), and they later describe this condition as requiring that accounts be judged in part on their "likelihood of motivating political action" (ibid., 404). Alexandre Gajevic Sayegh claims that "political traction" (2019, 121) is one of the criteria that theorists should consider when determining which use of the revenue generated by market-based instruments for emissions reduction they ought to endorse. What he seems to have in mind by this criterion is, at least roughly, how likely it is that a proposal will be taken seriously by policymakers and those in a position to influence policy.[18]

What Eric Posner and David Weisbach call "International Paretianism" (2010, 6) is probably the most well-known feasibility constraint that has been suggested as a constraint on our thinking about climate justice and policy. According to this constraint, any feasible climate treaty must be such that "all states believe themselves better off by their lights as a result of the climate treaty" (ibid., 6). Posner and Weisbach claim that because of this constraint, "feasibility rules out the vast redistributions of wealth that many believe are morally required" (ibid., 6). Instead, a feasible climate treaty would likely require side payments "from states that have a stronger interest in a climate treaty to states that have a weaker interest in a climate treaty" (ibid., 84). And since many states with an especially strong interest in a climate treaty are poor, and some with weaker interests are rich, "these side payments may go from poor to rich, at least in part" (ibid., 85).

It is worth noting an important difference between, on the one hand, the "incentive compatibility" constraint, and, on the other, the type of constraint represented by International Paretianism or more general concerns about political possibility. Brennan and Pettit's central claim seems to be that policymakers who are concerned to advance justice need to consider how the people who will be subject to whatever policies are adopted will in fact behave under different candidate policies. Because of this, theorists aiming to defend claims about what policies should be adopted, and who can be understood as (perhaps among other things) offering policy recommendations to legislators, must consider this as well. The way that considerations of feasibility constrain theorizing, then, is that claims that offer recommendations to policymakers must consider facts about the likely effects of selecting any particular policy from among the available options. These effects are, importantly, largely outside of the control of the policymakers themselves. Because of this, even policymakers who are motivated by nothing other than the aim of advancing justice must take the facts about likely effects into consideration when responsibly deciding which policies to try to enact. The "incentive compatibility" constraint on the advice that theorists ought to offer to policymakers, then, seems quite reasonable. The aim is simply to recommend the policies that would in fact be ethically optimal for legislators to enact, given what the effects of different candidate policies would actually be. There is nothing in the structure of this kind of constraint that requires the theorist to engage with those to whom normative recommendations are, in effect, being offered on any assumption other than that they are (or at least might be) as ethically motivated and committed to advancing justice as possible.[19]

International Paretianism and concerns about what can "generate political measures," or is "politically tractable," however, represent a different kind of constraint altogether. Theorists who operate within this kind of constraint cannot assume that those to whom their central normative claims offer recommendations are (or even might be) as ethically motivated as possible. Indeed, the central reason for accepting a constraint of this kind seems precisely that those people will likely not be sufficiently motivated by ethical considerations to do what it would be ethically best for them to do. Theorists who operate within this kind of constraint, then, cannot offer recommendations that constitute the morally best options for those to whom they are directed, in the way that those who operate within Brennan and Pettit's constraint can. Instead, they must aim, roughly, to offer recommendations that are ethically optimal *within the range of options that those to whom they are offered might actually choose*. This means that they must avoid recommending to decision-makers policies that would have ethically optimal effects if adopted, simply because those to whom the recommendations are to be offered will, predictably, not accept the recommendations.[20] This is a very different and, it seems to me, much more troubling constraint on how normative theorists ought to engage with those to whom they aim to offer normative guidance.

To see why this kind of constraint is particularly troubling, consider what it might imply about what a normative theorist ought to say to third parties about what, for example, policymakers ought to do with respect to climate change. It would, on the one hand, seem very strange if the constraint limited what the theorist ought to say, so that, for example, she should insist to her relatives, who are ordinary citizens, that it is not true, as a matter of justice, that legislators in the United States should be willing to ratify a climate treaty that would be costly to Americans overall, and to the well-off in particular. Moreover, it would seem objectionably inconsistent if the truth of the claim that justice requires that the United States sign on to an international treaty that would be costly overall somehow varied depending on to whom the claim was being directed.

In addition, if a feasibility constraint of the kind at issue were treated as constraining the *content* of requirements of justice, it would allow the mere unwillingness of some to behave in ways that are supported by strong moral reasons to render what would otherwise seem to be requirements of justice not requirements at all. In the case of climate justice, a feasibility constraint of this kind could imply that limiting warming to 2°C is not a requirement of justice. This is because there might be no set of climate policies that satisfies the following two conditions: (1) enough legislators could be motivated to support the policies for them to be enacted; and (2) enacting the policies would ensure that warming does not exceed 2°C. To take another example, Posner and Weisbach are explicit that International Paretianism likely provides theorists reason to recommend side payments from poor countries to rich countries as part of an international climate treaty (2010, 85). To the extent that this is true, taking International Paretianism as a constraint on the content of justice implies that imposing potentially significant costs of mitigation on poor countries is compatible with justice.

These are, it seems to me, quite implausible implications, if what we are aiming to do is to determine the content of the requirements of climate justice. This suggests that perhaps we should interpret the feasibility constraints in a different way. I suggest that they are best interpreted as proposed constraints on theorizing in urgent circumstances, such as those that we currently face with respect to climate change, rather than as constraints on the content of requirements of justice. Consider, for example, that Posner and Weisbach pursue their analysis of what an international climate treaty should look like within the bounds of International Paretianism, and endorse it as a constraint on this kind of theorizing, while claiming, first, that it "is not an ethical principle but a pragmatic constraint" (2010, 6), and second, that it is true, as a matter of ethical principle, that rich countries have an obligation to help poor ones (ibid., 74).

Assuming that I am right that the feasibility constraints that they defend cannot plausibly constrain the content of requirements of justice, why might theorists of climate justice nonetheless think that we should theorize within the constraints of what is politically tractable? One way of defending such a constraint on theorizing is to argue that the urgency of putting in place policies that will reduce emissions requires that theorists focus their attention only on potential measures that are feasible, in the sense of being likely enough to actually be adopted. Spending time reflecting on what an ideal of climate justice might require, when the proposals that result from such reflection have no chance of actually being implemented, is, it might be suggested, objectionably indulgent in conditions of such extreme urgency. This kind of philosophical reflection, it might be suggested, simply has nothing valuable to contribute to the necessary efforts to advance mitigation as quickly as possible, and these efforts require that everyone who is in a good position to contribute (including theorists working on climate justice) do so.

On this view, philosophical theorizing about climate justice that does not operate within the relevant feasibility constraints is simply idle, utopian intellectual activity that offers nothing of value in the fight against climate change. Perhaps this kind of intellectual work has more to offer when, for example, there is more time available to reflect on and debate a range of proposals for addressing injustice, or when failing to act right away will not ensure that later efforts will face much greater challenges, and may not be able to do more than limit the inevitable destruction. It might seem that when the need for immediate action is especially urgent, the role for ambitious philosophical theorizing simply vanishes. (For more on the role that justice theorists might play in policymaking, see McBee, J. and Kowarsch & Lenzi, both in this volume.)

V. THE ROLE OF AMBITIOUS THEORIZING IN URGENT CIRCUMSTANCES

While the aforementioned argument against ambitious theorizing in urgent circumstances can seem appealing, in my view it is mistaken. There are a number of important roles for theorizing that is not limited by the kinds of feasibility constraints that the argument advocates. And these roles can make it reasonable for theorists to engage in ambitious theorizing even if they are concerned that their work have the potential to contribute in a practical way to the fight against climate change. I will briefly highlight three important roles that such theorizing about climate justice can play.[21]

The first practical role that ambitious theorizing can play is contributing to making a greater range of policies or outcomes feasible over time. As Pablo Gilabert and Holly-Lawford Smith have emphasized, sometimes a policy or outcome that it is infeasible to bring about now can become feasible in the future, if we act in certain ways that are feasible now (Gilabert and Lawford-Smith 2012, 821–822; see also Gilabert 2009, 676–678; Gilabert 2011, 60–63; Gilabert 2012, 47–50; Lawford-Smith 2013, 249–250; Gilabert 2017). When this is the case, we may have "dynamic duties" (Gilabert 2009, 676–678) that require that we adopt a "transitional standpoint" (Gilabert and Lawford-Smith

2012, 821), aiming now to act in ways that will make it possible to satisfy a well-justified requirement at some point in the future. Arguments defending policies that are certainly not to be adopted in the near future or requirements that are certainly not to be met in virtue of near-term efforts can potentially contribute to making policies or outcomes that are infeasible now feasible in the future by encouraging reflection on ways that we might act now that could make it possible to bring them about later.[22] Insofar as there are strong moral reasons that support acting in ways that will make it feasible to bring such policies or outcomes about in the future, ambitious theorizing that generates reflection that makes such action more likely can contribute in a practical way to promoting justice over time.

In many cases, it is reasonable to take this as an aim in developing an argument about, for example, what is required as a matter of justice. And since consequential decisions about climate policy will need to be made not just in the near term, but for at least many decades in the future, there is no reason to think that this cannot be a reasonable aim of ambitious theorizing about climate justice.

A second important role for ambitious theorizing in urgent circumstances is that it can provide us with the intellectual basis necessary for holding accountable those who make it the case that certain policies that would significantly advance justice if adopted are infeasible. In the absence of ambitious theorizing, it might be easier for us, and for the public at large, to lose sight of the fact that there are important ethical questions that should be raised about the motivations, intentions, and behavior of those who make it the case that otherwise desirable policies are infeasible. Ambitious theorizing tends to encourage reflection on whether it is ethically acceptable for legislators, and many of the wealthy and powerful individuals and corporations that support them, to continue to make policies like (1)–(4) infeasible. When powerful agents make otherwise desirable policies infeasible, this in effect forces those who aim to advance climate justice to focus their attention on trying to get significantly less ambitious policies enacted. While this may be the right focus for those concerned about climate justice to adopt in those circumstances, it is important that we not lose sight of the fact that justice is, nonetheless, being undermined by others.

One valuable thing that theorizing that calls for negative judgments of the motivations, intentions, and behavior of those who help to make strong mitigation policies infeasible might do is contribute to motivating efforts among those who are concerned about climate justice and persuaded by the arguments that do not have as their main aim enacting the policies defended as requirements of justice. These efforts could aid mitigation efforts overall, despite not having any effect in terms of getting those policies enacted. (We have to assume that there is no chance that persuading concerned members of the public could have an effect on policy adoption, since if it could then we would, in effect, be suggesting that the feasible set is larger than we initially imagined—and if this is true then the argument against ambitious theorizing in urgent circumstances would not imply that the relevant arguments ought not be made). For example, an argument that suggests that taxes on the wealthy should be increased substantially in order to fund mitigation efforts might have no chance of making a difference to tax policy, since enough legislators and wealthy and influential citizens will oppose any such tax increase. But it might motivate some corporate leaders to adopt goals for their firms that include substantial emissions reductions and investments in clean energy technology, or motivate some individuals to invest in renewable energy firms rather than fossil-fuel companies. In other words, the policies advocated might themselves be infeasible in the sense that proponents of the argument against ambitious theorizing in urgent circumstances have in mind, but making the arguments could nonetheless still contribute to advancing the aims that explain why the policies are plausibly required as a matter of justice.

Where this is true, the value of making such arguments lies primarily in highlighting the values that support adopting the policies, which, at least in many cases, are directly relevant to justice. Because those values can often be promoted, at least to some extent, in ways other than via the policies, the arguments can highlight moral reasons for a range of actions other than those that have as their direct aim promoting the adoption of the policies. More generally, ambitious theorizing about justice can be valuable because a significant part of what it consists in is reflecting on which fundamental values are relevant to assessments of justice, and how those values are to be weighed up against one another and traded off in cases of conflict. This kind of reflection can aid moral decision-making in a wide variety of contexts apart from decisions about which public policies to support.[23]

In addition, there is independent value in providing the intellectual basis for holding accountable those who contribute to making otherwise desirable policies infeasible, even if doing this does not significantly affect either policy or how individuals behave in contexts in which justice-relevant values are implicated. Ambitious theorizing can provide grounds for justifiably concluding that those who stand in the way of policies that would promote climate justice are blameworthy, and this can generate justified indignation in those persuaded by the arguments that they might not otherwise have experienced. It might be objected that this kind of effect is not important enough to justify devoting valuable time to ambitious theorizing rather than to defending feasible policy proposals. Perhaps this is correct in cases in which this is the only valuable practical effect that making an ambitious argument might have. But I need not claim that this kind of effect is, by itself, sufficient to justify ambitious theorizing. I claim only that it is among a range of values that ambitious theorizing can serve that can together justify ambitious theorizing.

Finally, ambitious theorizing can help us to determine which option from a particular feasible set is best from the perspective of justice. This is not uncontroversial (see, e.g. Wiens 2015, 467–468), but in my view the skepticism that some have expressed is misplaced. I cannot argue for this claim in detail here, but in what follows I will note the central reason why I think that ambitious theorizing can provide valuable normative guidance in circumstances in which fully achieving justice is infeasible.

The most common argument for thinking that an account of what justice requires cannot helpfully guide action in conditions in which fully achieving justice is not feasible derives from the well-known "general theory of second best" (Lipsky and Lancaster 1956). The basic idea is that when we cannot satisfy all of some set of desiderata, it is not necessarily best to attempt to satisfy as many as possible, or, more generally, to approximate the ideal as much as possible. In discussions of justice and policy, this point is generally used in order to argue that we cannot move from the fact that a set of policies X is required as a matter of justice to the conclusion that when we cannot fully implement X, we should make our policy choices look as much like X as possible. In some cases, doing this will be worse in terms of the values that we have reason to care about.

The reason that this point, though undoubtedly correct, does not undermine the claim that ambitious theorizing can help to guide choice in conditions in which fully achieving justice is infeasible is that we need not, and in my view should not, understand the fundamental content of a theory of justice in terms of a set of policies. Instead, we should think that some of what justice requires is that we bring about certain outcomes (such as limiting warming to less than 2°C), and some of what it requires is that we weigh and trade-off justice-relevant values appropriately.

It is obvious that if we take limiting warming to 2°C as the content of a requirement of justice, we do not run into the second best problem in cases in which we will not in fact satisfy that requirement and must decide what to do among a range of feasible options. This is because we can choose on the basis of which option is most likely to minimize how far beyond the 2°C threshold we go.

Proponents of the argument against ambitious theorizing in urgent circumstances, however, may claim that any ambitious theorizing defending the 2°C requirement does not add anything to what would be recommended by the kind of theorizing that they favor, so that my point here does not support the claim that ambitious theorizing can play a valuable role that their preferred type cannot. This may be right in the case of the 2°C requirement, but, importantly, it is not true in more complex cases in which we must think about how to trade-off competing justice-relevant values, such as, for example, emissions reduction and short-term poverty reduction. Ambitious theorizing requires determining, as a matter of principle, how competing values should ideally be weighed against each other. Theorizing within the kinds of feasibility constraints endorsed by proponents of the argument against ambitious theorizing in urgent circumstances, on the other hand, generally does not involve reflecting on fundamental questions of this kind.[24] In addition, only theorizing in the absence of feasibility constraints will involve reflecting on how these trade-offs would *ideally* be made, when all of the options are considered on the table. Only ambitious theorizing, then, can generate principles that can serve as appropriate guides for making these decisions across all possible cases.

There are, then, a number of important roles that can be played by ambitious philosophical theorizing about climate justice, even in our current urgent circumstances. While it can be tempting to think that work on climate justice ought to be focused more directly on advocating feasible measures to fight climate change, I have argued that this view involves an unduly narrow conception of the ways that philosophical work can contribute to efforts to promote justice in our admittedly very non-ideal world.[25]

NOTES

1. Perhaps most prominently, there has been much criticism of "ideal theory" in political philosophy. See, for example, Mills (2005); Sen (2009); Gaus (2016).

2. Ambitious theorizing about climate justice might also be defended by appeal to the claim that knowledge about important matters is valuable, even if possessing it will serve no further practical purpose (Estlund 2011a, 2014, 132–134; see also Räikkä 1998, 30; Cohen 2008, 268). Although I find this claim plausible, in this chapter I will rely only on practical roles that ambitious theorizing can, in my view, play.

3. It should go without saying that my defense of the view that there are valuable roles for ambitious theorizing in the struggle against climate change is consistent with recognizing the important contributions that can be made by normative philosophical work that focuses on what can be done to improve things in terms of climate justice given constraints such as the motivations and behavioral tendencies of those who will not do what ambitious theories might imply that they ought to do. For just a few recent examples of valuable contributions of this kind, see Moellendorf (2015, 2016); Caney (2016); Roser (2016). It is worth noting that all three of these theorists have also contributed to what I have called ambitious theorizing about climate justice (see, e.g., Caney (2005, 2010, 2012); Moellendorf (2014); Meyer and Roser (2006, 2010)).

4. For an influential account of why it has been so difficult to make progress in addressing the threat of climate change, despite the great moral importance of doing so, see Gardiner (2011).

5. A recent IPCC Special Report estimates that limiting warming to below 2°C would require GHG emissions reductions of 25 percent from 2010 levels by 2030, and net zero emissions by 2070 (IPCC 2018, 12). Achieving the more ambitious 1.5°C limit is estimated to require a 45 percent reduction in emissions from 2010 levels by 2030, and net zero emissions by 2050 (IPCC 2018, 12).

6. For discussion of the demandingness of our climate change-related obligations, see Fruh and Hedahl (2013); Berkey (2014); Fragnière (2018).

7. A recent study with a conclusion toward the higher end of current estimates suggests that the global social cost of carbon is roughly $417 per ton of CO_2, while also providing estimates of the differential costs at the country level (Ricke et al. 2018). In contrast, William Nordhaus (2017) estimates that the social cost of carbon is roughly $31 per ton.

8. The International Energy Agency has concluded that "no more than one-third of proven reserves of fossil fuels can be consumed prior to 2050 if the world is to achieve the 2°C goal" (IEA 2012; see also McGlade and Ekins 2015).

9. This could be done in a variety of ways, for example via any combination of: (1) a significantly increased marginal income tax rate on earnings over a certain rather high amount; (2) an increase in the estate tax; (3) an increase in the capital gains tax rate; (4) a wealth tax on individuals whose net worth exceeds a certain rather high amount; (5) an increase in the corporate tax rate.

10. This could be either because they would not comply with the explicit requirements of the policies, or because they would behave in ways that are permitted under the policies, but that would nonetheless make it the case that the goals that the policies were intended to promote would not in fact be advanced. For discussion of what we should say about what ought to be done in cases with the latter structure, see Southwood (2016).

11. As Juha Räikkä points out (1998, 28–31), it is widely accepted that any feasibility condition in political theory (that is, any feasibility condition on requirements of justice) cannot rule out as infeasible any policy that cannot be implemented quickly due to the opposition of powerful political groups (even if those groups oppose the policy on moral grounds rather than for self-interested reasons). Such policies might be "politically infeasible," but they do not, just in virtue of that fact, fail whatever feasibility condition might constrain the content of requirements of justice.

12. The conclusion that the best feasible mitigation policies would involve compensation to well-off high emitters, paid either by the governments of nations most vulnerable to climate change, many of which are poor, or by future generations, is endorsed by a number of scholars (e.g., Posner and Sunstein 2008; Posner and Weisbach 2010; Broome 2012; for discussion see Budolfson forthcoming)

13. Studies that aim to estimate the emissions reductions that are necessary in order to keep warming below a particular limit typically state their conclusions in terms of what will provide either a 50% or a 66% probability of sufficiently limiting warming. For analyses using the 2°C target, see; Friedlingstein et al. (2014); Raupach et al. (2014); McGlade and Ekins (2015); Heede and Oreskes (2016); Rekker et al. (2018). For analyses using the 1.5°C target, see Millar et al. (2017); Matthews et al.

(2017); Goodwin et al. (2018); Kriegler et al. (2018); Leach et al. (2018); Rogelj et al. (2018); Tokarska and Gillett (2018).

14. In making this assumption, I am also assuming that at least a subset of the requirements of climate justice are outcome-focused, at least in a broad sense, since satisfying the requirements consists in ensuring that a certain state of affairs is brought about (e.g., one in which warming is limited to less than 2°C (or 1.5°C), or one in which emissions are limited to less than a particular amount). This is consistent with the way that at least most philosophers working on climate justice seem to conceive of these requirements, including those who think that other components of climate justice, such as the appropriate distribution of mitigation burdens, can plausibly be understood as matters of pure procedural justice (e.g., Brandstedt and Brulde 2019, 796).

15. Geoffrey Brennan is explicit that, in his view, the "ought implies can" principle is a "bad point of departure for discussion of feasibility considerations" (2013, 322). He argues that the conception of feasibility that political philosophers work with should be expanded "in ways that would be hospitable to the incorporation of 'economic and political'" considerations (ibid., 315).

16. Brennan endorses the common complaint made by economists that philosophers do not take these kinds of feasibility considerations sufficiently seriously (2013, 328). For criticism of the claim that work in political philosophy should, as a general matter, be pursued within constraints set by current social scientific findings, see McTernan (2019). For criticism of the view that political philosophy should be pursued within constraints set by assumptions about human motivation commonly accepted by economists, see Herzog (2015).

17. If the normative enterprise in question is determining what policies should be adopted in the actual world, then this claim seems right. But philosophers who reject the view that compliance constraints and incentive compatibility should be regarded as constraints on the content of requirements of justice do not take this as the central aim of their theorizing. It is unclear, then, exactly who Brennan and Pettit think make the mistake of recommending the adoption of policies in the actual world that would predictably fail to advance justice as much as alternative policies (or are committed by their theorizing to making such recommendations).

18. Gajevic Sayegh also describes his methodological approach to thinking about climate justice as requiring that "imperatives and recommendations from other disciplines" (including, uncontroversially in my view, climate science, but also economics) "constrain the philosophical analysis" (2019, 114). For a critical discussion of Gajevic Sayegh's view, see Berkey (2019).

19. In my view, this does not, however, provide any reason to accept that incentive compatibility is a constraint on the content of requirements of justice. This is because while it is reasonable to advise policymakers about what they ought to do in light of the best available evidence about how people subject to the policies will actually behave, we should not allow anyone's potentially objectionable behavioral tendencies, intentions, and attitudes to constrain the content of justice. Ambitious theorizing, then, need not be limited by the concerns about incentive compatibility highlighted by Brennan and Pettit, even though there are certainly valuable contributions to be made by theorizing that does take incentive compatibility as a constraint.

20. This, presumably, is the way that we must understand the claim that policies such as (1)–(4) are infeasible, since it seems likely that, conditional on actually being adopted, their effects would be quite good from an ethical perspective.

21. The roles that I describe here are not the only ones that I believe ambitious theorizing can play in urgent circumstances. I cannot, however, offer a more comprehensive discussion in this chapter.

22. Of course, the extent to which such arguments will tend to encourage reflection that otherwise would not have occurred, and thereby potentially make a difference to the feasibility of policies over time, will depend at least to some extent on to whom the arguments are directed, and on who actually engages with them. It may be that ambitious theorizing aimed at an audience broader than academic philosophers has much more potential to contribute in this way than work aimed purely at professional philosophers. If this is right, then one interesting upshot of my argument may be that it calls for philosophers to engage in more ambitious theorizing aimed at a broad audience.

23. It is worth noting that a further feature of ambitious theorizing that we have good reason to value is that it, and its recommendations, are capable of being addressed to all agents who are in a position to promote justice, since it does not require tailoring the content of theoretical claims and policy recommendations to the audience being addressed. When a theorist operates within feasibility constraints of the kind endorsed by proponents of the argument against ambitious theorizing in urgent circumstances, however, what they ought to recommend to different audiences will vary, because what different agents might be willing to do will vary.

24. This kind of theorizing, of course, will typically involve considering certain kinds of trade-offs. Posner and Weisbach, for example, suggest that emissions reduction should be prioritized over the near-term redistribution of resources to the global poor because they believe that in the long run emissions reduction would have a greater net positive effect on human welfare. This, however, is not a trade-off between competing fundamental values, but instead a question about how to best promote a single value in non-ideal conditions.

25. I am grateful to the audience at the 2020 Rocky Mountain Ethics Congress, at which this chapter was presented. Chetan Cetty provided very helpful comments for that session. I also received helpful written comments from Corey Katz, Sarah Kenehan, and Kian Mintz-Woo. In addition, I have benefitted from discussion with Alyssa Bernstein.

REFERENCES

Alliance of Small Island States. 2019. "About Us." https://www.aosis.org/about/.
Berkey, Brian. 2014. "Climate Change, Moral Intuitions, and Moral Demandingness." *Philosophy and Public Issues* 4(2): 157–189.
Berkey, Brian. 2019. "Climate Justice, Climate Policy, and the Role of Political Philosophy." *Ethics, Policy, & Environment* 22(2): 145–147.

Brandstedt, Eric, and Anna-Karin Bergman. 2013. "Climate Rights: Feasible or Not?" *Environmental Politics* 22(3): 394–409.

Brandstedt, Eric, and Bengt Brulde. 2019. "Towards a Theory of Pure Procedural Climate Justice." *Journal of Applied Philosophy* 36(5) (November): 785–799.

Brennan, Geoffrey. 2013. "Feasibility in Optimizing Ethics." *Social Philosophy & Policy* 30(1–2) (January): 314–329.

Brennan, Geoffrey, and Philip Pettit. 2007. "The Feasibility Issue." In *The Oxford Handbook of Contemporary Philosophy*, edited by Frank Jackson and Michael Smith, 258–279. Oxford: Oxford University Press.

Broome, John. 2012. *Climate Matters: Ethics in a Warming World*. New York: W.W. Norton & Company.

Budolfson, Mark. Forthcoming. "Political Realism, Feasibility Wedges, and Opportunities for Collective Action on Climate Change." In *Philosophy and Climate Change*, edited by Mark Budolfson, Tristram McPherson, and David Plunkett. Oxford: Oxford University Press.

Caney, Simon. 2005. "Cosmopolitan Justice, Responsibility, and Global Climate Change." *Leiden Journal of International Law* 18(4) (December): 747–775.

Caney, Simon. 2010. "Climate Change and the Duties of the Advantaged." *Critical Review of International Social and Political Philosophy* 13(1): 203–228.

Caney, Simon. 2012. "Just Emissions." *Philosophy & Public Affairs* 40(4) (Fall): 255–300.

Caney, Simon. 2016. "Climate Change and Non-Ideal Theory: Six Ways of Responding to Non-Compliance." In *Climate Justice in a Non-Ideal World*, edited by Clare Heyward and Dominic Roser, 21–42. Oxford: Oxford University Press.

Chahboun, Naima. 2017. "Three Feasibility Constraints on the Concept of Justice." *Res Publica* 23(4) (November): 431–452.

Cohen, G.A. 2008. *Rescuing Justice and Equality*. Cambridge: Harvard University Press.

Estlund, David. 2011a. "What Good Is It? Unrealistic Political Theory and the Value of Intellectual Work." *Analyse & Kritik* 33(2) (November): 395–416.

Estlund, David. 2011b. "Human Nature and the Limits (If Any) of Political Philosophy." *Philosophy & Public Affairs* 39(3) (Summer): 207–237.

Estlund, David. 2014. "Utopophobia." *Philosophy & Public Affairs* 42(2) (Spring): 113–134.

Estlund, David. 2016. "Reply to Wiens." *European Journal of Political Theory* 15(3): 353–362.

Fragnière, Augustin. 2018. "How Demanding is Our Climate Duty? An Application of the No-Harm Principle to Individual Emissions." *Environmental Values* 27(6) (December): 645–663.

Friedlingstein, Pierre, Robbie M. Andrew, Joeri Rogelj, Glen P. Peters, Josep G. Canadell, Reto Knutti, Gunnar Luderer, et al. 2014. "Persistent Growth of CO_2 Emissions and Implications for Reaching Climate Targets." *Nature Geoscience* 7(10) (October): 709–715.

Fruh, Kyle, and Marcus Hedahl. 2013. "Coping with Climate Change: What Justice Demands of Surfers, Mormons, and the Rest of Us." *Ethics, Policy, & Environment* 16(3): 273–296.

Gajevic Sayegh, Alexandre. 2019. "Pricing Carbon for Climate Justice." *Ethics, Policy, & Environment* 22(2): 109–130.

Gardiner, Stephen M. 2011. *A Perfect Moral Storm: The Ethical Tragedy of Climate Change*. New York: Oxford University Press.

Gaus, Gerald. 2016. *The Tyranny of the Ideal: Justice in a Diverse Society*. Princeton: Princeton University Press.

Gheaus, Anca. 2013. "The Feasibility Constraint on the Concept of Justice." *Philosophical Quarterly* 63(252) (July): 445–464.

Gilabert, Pablo. 2009. "The Feasibility of Basic Socioeconomic Human Rights: A Conceptual Exploration." *Philosophical Quarterly* 59(237) (October): 659–681.

Gilabert, Pablo. 2011. "Feasibility and Socialism." *Journal of Political Philosophy* 19(1) (March): 52–63.

Gilabert, Pablo. 2012. "Comparative Assessments of Justice, Political Feasibility, and Ideal Theory." *Ethical Theory and Moral Practice* 15(1) (February): 39–56.

Gilabert, Pablo. 2017. "Justice and Feasibility: A Dynamic Approach." In *Political Utopias: Contemporary Debates*, edited by Kevin Vallier and Michael Weber, 95–126. New York: Oxford University Press.

Gilabert, Pablo, and Holly Lawford-Smith. 2012. "Political Feasibility: A Conceptual Exploration." *Political Studies* 60(4) (December): 809–825.

Goodwin, Philip, Anna Katavouta, Vassil M. Roussenov, Gavin L. Foster, Eelco J. Rohling, and Richard G. Williams. 2018. "Pathways to 1.5°C and 2°C Warming Based on Observational and Geological Constraints." *Nature Geoscience* 11(2) (February): 102–107.

Hamlin, Alan. 2017. "Feasibility Four Ways." *Social Philosophy & Policy* 34(1) (Summer): 209–231.

Heede, Richard, and Naomi Oreskes. 2016. "Potential Emissions of CO_2 and Methane from Proved Reserves of Fossil Fuels: An Alternative Analysis." *Global Environmental Change* 36 (January): 12–20.

Herzog, Lisa. 2015. "Distributive Justice, Feasibility Gridlocks, and the Harmfulness of Economic Ideology." *Ethical Theory and Moral Practice* 18(5) (November): 957–969.

IEA (International Energy Agency). 2012. *World Energy Outlook 2012*. Paris.

IPCC (Intergovernmental Panel on Climate Change). 1990. *Climate Change: The IPCC Scientific Assessment*, edited by J. T. Houghton, G. J. Jenkins, and J. J. Ephraums. Cambridge: Cambridge University Press.

IPCC (Intergovernmental Panel on Climate Change). 2018. "Summary for Policymakers." In *Global Warming of 1.5°C: An IPCC Special Report on the Impacts of Global Warming of 1.5°C Above Pre-Industrial Levels and Related Global Greenhouse Gas Emission Pathways, in the Context of Strengthening the Global Response to the Threat of Climate Change, Sustainable Development, and Efforts to Eradicate Poverty*, edited by Valérie Masson-Delmotte, Panmao Zhai, Hans-Otto Pörtner, Debra Roberts, Jim Skea, Priyadarshi R. Shukla, Anna Pirani, et al. In Press.

Kriegler, Elmar, Gunnar Luderer, Nico Bauer, Lavinia Baumstark, Shinichiro Fujimori, Alexander Popp, Joeri Rogelj, Jessica Strefler, and Detlef P. van Vuuren. 2018. "Pathways Limiting Warming to 1.5°C: A Tale of Turning Around in No

Time?" *Philosophical Transactions of the Royal Society A: Mathematical, Physical and Engineering Sciences* 376(2219) (May): 1–17.

Lawford-Smith, Holly. 2013. "Understanding Political Feasibility." *Journal of Political Philosophy* 21(3) (September): 243–259.

Leach, Nicholas J, Richard J. Millar, Karsten Haustein, Stuart Jenkins, Euan Graham, and Myles R. Allen. 2018. "Current Level and Rate of Warming Determine Emissions Budgets under Ambitious Mitigation." *Nature Geoscience* 11(8) (August): 574–579.

Lipsky, R. G., and Kelvin Lancaster. 1956. "The General Theory of Second Best." *Review of Economic Studies* 24(1): 11–32.

Matthews, H. Damon, Jean-Sébastien Landry, Antti-Ilari Partanen, Myles Allen, Michael Eby, Piers M. Forster, Pierre Friedlingstein, and Kirsten Zickfeld. 2017. "Estimating Carbon Budgets for Ambitious Climate Targets." *Current Climate Change Reports* 3(1) (March): 69–77.

McGlade, Christope, and Paul Ekins. 2015. "The Geographical Distribution of Fossil Fuels Unused When Limiting Global Warming to 2°C." *Nature* 517(7533) (January): 187–190.

McTernan, Emily. 2019. "Justice, Feasibility, and Social Science as It Is." *Ethical Theory and Moral Practice* 22(1) (February): 27–40.

Meyer, Lukas H., and Dominic Roser. 2006. "Distributive Justice and Climate Change: The Allocation of Emission Rights." *Analyse & Kritik* 28(2) (November): 223–249.

Meyer, Lukas H., and Dominic Roser. 2010. "Climate Justice and Historical Emissions." *Critical Review of International Social and Political Philosophy* 13(1): 229–253.

Millar, Richard J., Jan S. Fuglestvedt, Pierre Friedlingstein, Joeri Rogelj, Michael J. Grubb, H. Damon Matthews, Ragnhild B. Skeie, Piers M. Forster, David J. Frame, and Myles R. Allen. 2017. "Emission Budgets and Pathways Consistent with Limiting Warming to 1.5°C." *Nature Geoscience* 10(10) (October): 741–747.

Mills, Charles W. 2005. "'Ideal Theory' as Ideology." *Hypatia* 20(3) (Summer): 165–184.

Moellendorf, Darrel. 2014. *The Moral Challenge of Dangerous Climate Change: Values, Poverty, and Policy*. Cambridge: Cambridge University Press.

Moellendorf, Darrel. 2015. "Can Dangerous Climate Change Be Avoided?" *Global Justice: Theory Practice Rhetoric* 8(2): 66–85.

Moellendorf, Darrel. 2016. "Taking UNFCCC Norms Seriously." In *Climate Justice in a Non-Ideal World*, edited by Clare Heyward and Dominic Roser, 104–121. Oxford: Oxford University Press.

Nordhaus, William D. 2017. "Revisiting the Social Cost of Carbon." *PNAS* 114(7): 1518–1523.

Posner, Eric, and Cass R. Sunstein. 2008. "Climate Change Justice." *Georgetown Law Journal* 96: 1565–1612.

Posner, Eric A., and David Weisbach. 2010. *Climate Change Justice*. Princeton: Princeton University Press.

Räikkä, Juha. 1998. "The Feasibility Condition in Political Theory." *Journal of Political Philosophy* 6(1) (March): 27–40.

Raupach, Michael R., Steven J. Davis, Glen P. Peters, Robbie M. Andrew, Josep G. Canadell, Philippe Ciais, Pierre Friedlingstein, Frank Jotso, Detlef P. van Vuuren, and Corinne Le Quéré. 2014. "Sharing a Quota on Cumulative Carbon Emissions." *Nature Climate Change* 4(10) (October): 873–879.

Rekker, Saphira A. C., Katherine R. O'Brien, Jacquelyn E. Humphrey, and Andrew C. Pascale. 2018. "Comparing Extraction Rates of Fossil Fuel Producers Against Global Climate Goals." *Nature Climate Change* 8(6) (June): 489–492.

Ricke, Katharine, Laurent Drouet, Ken Caldeira, and Massimo Tavoni. 2018. "Country-Level Social Cost of Carbon." *Nature Climate Change* 8(10) (October): 895–900.

Rogelj, Joeri, Alexander Popp, Katherine V. Calvin, Gunnar Luderer, Johannes Emmerling, David Gernat, Shinichiro Fujimora, et al. 2018. "Scenarios Toward Limiting Global Mean Temperature Increase Below 1.5°C." *Nature Climate Change* 8(4) (April): 325–332.

Roser, Dominic. 2016. "Reducing Injustice Within the Bounds of Motivation." In *Climate Justice in a Non-Ideal World*, edited by Clare Heyward and Dominic Roser, 83–103. Oxford: Oxford University Press.

Sen, Amartya. 2009. *The Idea of Justice*. Cambridge: Harvard University Press.

Southwood, Nicholas. 2016. "Does 'Ought' Imply 'Feasible'?" *Philosophy & Public Affairs* 44(1) (Winter): 7–45.

Tokarska, Katarzyna B., and Nathan P. Gillett. 2018. "Cumulative Carbon Emissions Budgets Consistent with 1.5°C Global Warming." *Nature Climate Change* 8(4) (April): 296–301.

United States Department of Defense. 2015. "National Security Implications of Climate-Related Risks and a Changing Climate." https://archive.defense.gov/pubs/150724-congressional-report-on-national-implications-of-climate-change.pdf?source=govdelivery.

Wallace-Wells, David. 2019. *The Uninhabitable Earth: Life After Warming*. New York: Penguin Random House.

Wiens, David. 2014. "'Going Evaluative' to Save Justice from Feasibility—A Pyrrhic Victory." *Philosophical Quarterly* 64(255) (April): 301–307.

Wiens, David. 2015. "Political Ideals and the Feasibility Frontier." *Economics and Philosophy* 31(3) (November): 447–477.

Wiens, David. 2016. "Motivational Limitations on the Demands of Justice." *European Journal of Political Theory* 15(3): 333–352.

Working Groups I, II, and III to the Fifth Assessment Report of the Intergovernmental Panel on Climate Change. 2015. *Climate Change 2014: Synthesis Report*. Geneva, CH: IPCC.

Chapter 5

Is a Just Climate Policy Feasible?

Kirsten Meyer

Industrialized countries have failed to significantly reduce their emissions of greenhouse gases (GHGs). These high emissions violate demands of global and intergenerational justice.[1] Moreover, they will not reach the self-imposed emissions reduction targets they committed to in accord with the Paris Agreement. This failure to comply with the demands of climate justice is often explained by reference to feasibility constraints. For example, some have argued that there is little evidence that global action at the scale and speed necessary to limit warming to 1.5°C is feasible (Jewell and Cherp 2020).

An outcome is feasible if it is possible for an agent to bring it about if he or she tries. It is infeasible if it is not possible to bring it about, no matter what an agent might try to do. In addition to this binary notion, feasibility can be understood as a matter of degree (Gilabert and Lawford-Smith 2012; Lawford-Smith 2013). Certain constraints, such as logical, nomological, and biological constraints, can be considered "hard" in the sense that they can make an action fully infeasible. "Soft" constraints, such as economic, institutional, and cultural circumstances, as well as other people's lack of motivation, make an outcome less feasible but not infeasible. In this chapter, I will focus primarily on soft constraints, which make an outcome less feasible.[2]

Critics of ideal theory often point to these constraints. They claim that ideal theory neglects feasibility considerations and is therefore not helpful in formulating realistic policy proposals. However, I will argue that ideal theory can shed a valuable light on these constraints. More specifically, I will argue that feasibility constraints can sometimes result from a lack of justice. Because ideal theory helps us identify standards of justice, it can relevantly contribute to the formulation of policies that are less affected by certain feasibility constraints.

In the first section, I will discuss different kinds of feasibility constraints on the reduction of GHG emissions and spell out the kind of feasibility constraints upon which I will focus. Next, I will analyze feasibility constraints on carbon pricing specifically because these will helpfully illustrate the thesis of this chapter: feasibility constraints often reflect, or even stem from, a lack of social justice. Theoretically, it could also be the other way around: unjust policies might sometimes be more feasible. I will argue, however, that in these cases, feasibility would not be an appropriate goal. I will then elaborate on the connection between feasibility and non-ideal theory. Ideal theory is sometimes accused of being irrelevant for actual political decisions since it neglects feasibility constraints. In contrast, I will argue that instead of neglecting feasibility constraints, ideal theory has the potential to explain some of them. By explaining some of them, it paves the way for more just and thus (hopefully) for more feasible climate policies.

I. FEASIBILITY CONSTRAINTS AND CLIMATE ACTION

To begin, I will shed light on the most relevant aspects of feasibility regarding the reduction of GHGs and distinguish between different levels of political feasibility.

There are various feasibility constraints that hinder our ability to significantly reduce GHGs. First of all, we have to acknowledge technical constraints. For example, technological constraints currently limit the development of long-distance airplanes that are not powered by fossil fuels. Physical constraints are those created by scientific laws, systems and behaviors, such as the behavior of the Earth's climate. Physical constraints that cannot be overcome by any technology can count as "hard" constraints. "Soft" constraints are more important, however, because they are the ones we can attempt to change.

Among these, there are constraints on the potential for implementing certain policies at the international level. For example, there are constraints on the implementation of policies that would reduce the number of flights. Although road fuel is taxed, aviation kerosene, which is used in jet engines, is exempt from tax. Taxing aviation fuel, however, is prevented by international treaties that a single government cannot overcome.[3] Another example are constraints that restrict the ability of the EU and its member states to adopt policies that reduce their GHG emissions. EU regulations constrain unilateral efforts by member states. Conversely, the national interests of individual member states can also stand in the way of stricter EU regulations aimed at environmental and climate protection.

In contrast to hard, or physical constraints, in principle these constraints could be overcome. A tax on aviation fuel could be achieved by international political agreements,[4] or EU member states could work together to enact a more ambitious policy if all parties involved had the political will to do so. However, a lack of this political will by some parties could very well be described as a feasibility constraint in the sense that the relevant norms would be prevented from being implemented at the international level.

Political constraints at the national level are another example of a relevant soft constraint, and these will be the focus of my essay. Specifically, electoral support for a political measure functions as a feasibility constraint in democratic countries. Even if the government could introduce an ambitious climate policy (such as a high price on carbon), if it is against the general will of the electorate, such a policy might be unlikely to survive over the long run and so count as infeasible for that reason (cf. Lawford-Smith 2013, 252).

Why are certain policy proposals politically infeasible in this way? First of all, the interests of powerful lobbying groups influence politics. Moreover, individual voters may understand the situation in ways that limit their motivation to support a particular policy. For example, even though the high personal emissions of people living in the industrialized countries are unjust,[5] it seems that people are not willing to reduce their carbon footprint. Some people may underestimate their contribution to climate change—perhaps even due to a lack of knowledge. Others might think that their high emissions are not unjust. Moreover, even if they do think that their emissions are unjust, they could still lack the motivation to reduce them. They may, for example, assume that justice demands too much of them if it asks them to significantly change their lifestyle. All of these reasons may contribute to their unwillingness to endorse policy proposals (such as a high tax on carbon) that would force them to reduce their emissions.

How relevant are these kind of feasibility constraints? How important is it to overcome a lack of electoral support for certain climate policies? Simon Caney points out that climate policy needs "sufficient electoral support for governments to implement radical mitigation policies and sustain them over time" (Caney 2016, 21). According to Caney, this is a difficult endeavor because one has to communicate the effects of climate change in such a way that it resonates with citizens. But this is a challenging task: the protection of the climate is very demanding, and people have to change their habits quite radically—more than most are willing to. For this reason, ambitious proposals for climate policy risk the resistance of some of those who would be affected by them, and who would have to adjust their behavior as a result of the policy.

Consider, for instance, the so-called yellow vests movement in France, which began in October 2018. Among other factors, the movement was sparked by rising fuel prices, and it reflected a resistance to a newly implemented carbon tax. However, it should be noted that the protest movement was not so much opposed to the increased tax on gas and fuel as such, but rather to its unjust implementation as it was planned by the French government. In fact, one could make the case that concerns of social justice were the primary motivation of the protest. Participants appear to be characterized by modest income and demands for more redistribution (Gagnebin et al. 2019, 5). In 2019, fearing similar protests, the German government refrained from adopting a higher price on carbon. Simply put, even though a higher price on carbon would have been more just from a global or intergenerational perspective, this policy was assumed to be practically infeasible.

In the following section, I will elaborate more on this case as it exemplifies the thesis of this chapter: feasibility constraints can sometimes result from a lack of justice, and climate ethics is a valuable tool for explaining them.

II. FEASIBILITY CONSTRAINTS ON CARBON PRICING

Feasibility constraints are often framed as constraints on the implementation of an (ideally) just climate policy. However, constraints on climate policy may also result from a lack of justice *within* certain climate policy proposals. In order to demonstrate this, I will focus on a concrete policy proposal: a price on carbon in Germany. Putting a price on carbon is among the most important climate policies, and in Germany, the political discourse is heavily focused on market-based measures like emissions trading, which is one form of carbon pricing.

Germany, like all EU member states, participates in the European Union Emissions Trading System (EU ETS). Until now, however, emissions caused by the transportation and building heating sectors have been exempt from an EU- or Germany-wide price. Both sectors rely heavily on fossil fuels, such as heating oil, natural gas, gasoline, and diesel. Due to this, in 2019, Germany proposed a national emissions pricing system for transportation and heating emissions parallel to the EU-wide system in order to cover important GHG emissions not included in the ETS.

In September 2019, the German government, supported by a coalition of the German Christian Democratic Union (CDU), Christian Social Union (CSU) und Social Democratic Party of Germany (SPD), agreed on an outline for a climate plan meant to ensure that the country reaches its emission reduction targets in line with the Paris Agreement. More specifically, they agreed Germany would introduce a national carbon pricing system in the transportation and building sectors. The national carbon pricing would start in 2021. Originally, the Cabinet had proposed a graduated price ranging from 10 euros in 2021 to 35 euros per ton of CO_2 equivalents by 2025. This was heavily criticized by experts, business associations, and environmental NGOs. The low entry-level price in 2021 and the progression in following years were expected to have hardly any steering effect. In the course of the negotiations in the Mediation Committee of Bundestag and Bundesrat, it was decided that the price should be increased.[6] According to the compromise of the Mediation Committee, the fixed price for 2021 will be 25 euros per ton of CO_2 equivalents, and it will rise from then on. Even this increased price, however, has been criticized for still being too low to be effective.

What is the reasoning behind setting both the starting prices and the incremental increases at levels so low that they might be ineffective? In defending her rather moderate proposal, Chancellor Angela Merkel said that "politics is what is possible" (Frankfurter Allgemeine Zeitung, 2019). As one can infer from this statement, a higher price on carbon was assumed to be politically infeasible. This alleged fact was often explained as follows: too many people do not obey what climate justice demands and do not want to change their lifestyle, and so, their reluctance requires a deviation from what ideal justice demands.

In contrast to this line of argument, however, I understand the relevant feasibility constraints differently. I argue that the feasibility constraints on a higher price on carbon mentioned by Chancellor Merkel do not pose a problem *for* climate justice. Instead, these constraints reflect a lack *of* climate justice. The example of carbon pricing reveals that sometimes a policy may not be feasible *because* it is unjust.

In the policy proposal under discussion, a lack of justice stems from the fact that those with lower incomes bear a greater burden by having to pay a higher price on carbon. In contrast, those who are more affluent can more easily afford to pay the price on carbon or for activities that are more expensive because of that price. Indeed, Tank (2020) argues that all politically relevant forms of carbon pricing should be considered unfair. To begin with, Tank points out why carbon mitigation policies burden those who are less affluent to a greater degree. First, many less affluent people will have to cut down on their car travel as a result of robust carbon pricing (due to the higher fuel prices), while other individuals will be able to afford the changes demanded by the policy. And even if the more affluent will need to change their behavior as well, they will usually have a greater selection of comparable substitutes open to them. For example, even if a more affluent person can no longer afford to drive her emissions-intensive car, she is more likely to be able to substitute it with an electric car (Tank 2020, 5).

Furthermore, Tank points out that placing greater burdens on the less affluent than on those who are more affluent is in conflict with plausible moral principles for the distribution of burdens in the context of climate change mitigation. According to the Polluter Pays Principle (PPP), as Tank interprets it, agents should bear the burdens of addressing a problem in proportion to their contribution to causing the problem. Since the vast majority of more affluent individuals emit more CO_2, and thus bear a greater responsibility for the problem of climate change than less affluent individuals, any mitigation policy that burdens the more affluent less than the less affluent seems unfair. Thus, carbon pricing, as proposed by the German plan, is in conflict with the PPP. Moreover, Tank points out that the "Ability to Pay Principle" (APP) also supports the claim that carbon pricing is unfair since the less affluent have less ability to afford higher costs associated with a price on carbon (Tank 2020, 7–10).

Are there forms of carbon pricing that manage to be fair in that they avoid imposing heavier burdens on the less affluent than on the wealthy? According to Tank, if significant parts of society were exempted from having to make behavioral changes this will mean that overall emissions will not sufficiently be reduced, even if those parts of society emit relatively little GHGs. He thinks that this is also troubling for those policies that offer a full (or near-full) refund for less affluent individuals: "Using some of the revenue from carbon pricing to refund less affluent people can play a role in a fair and effective carbon pricing scheme, but too-generous refunds might threaten its effectiveness if they allow too many individuals to keep on emitting like before" (Tank 2020, 12).

Thus, Tank argues that *if* the system of carbon pricing is *fair*, it will also be *ineffective*. However, I do not think that this is necessarily the case. Even a system with generous refunds could nevertheless prevent people from emitting if relevant and attractive alternatives, such as frequent and free public transport, are available. Subsidized public transport could become cheaper for every person than taking the car even without an additional price on carbon. A price on carbon can nevertheless be a further incentive to take the bus. In addition, even if the less affluent are compensated according to a fair pricing scheme, this does not mean that they will use the money to continue to drive their cars. This money could be spent on other things that do not contribute to increased emissions. Creating these kind of incentives does not seem to be unfair.

Tank also formulates a further feasibility constraint. He thinks that a fair system is *politically* infeasible (Tank 2020, 12). We should, however, be more confident than Tank in this regard. One could suggest what a *fair* price on carbon might be, argue *why* it is fair, and then hope that the political feasibility constraint diminishes. Indeed, in Germany, the Green Party did suggest a complete per capita reimbursement of a high price on carbon. In contrast, the coalition argued for a lower price on carbon and thereby hinted at feasibility constraints. Politicians of the SPD might have been worried that people with lower incomes will no longer support the coalition if the tax is too high. Politicians of the CDU/CSU might have worried that they would lose the more affluent voters' support if the proposal contained a large transfer of wealth. The result was a low price on carbon without an ambitious form of reimbursement. Thus, it turned out to be ineffective as well as unfair. Instead, a higher price on carbon with an equal per capita reimbursement, as suggested by the Green Party, would have been both more effective and more fair.

The actual carbon pricing model that was adopted burdens poor households more than rich households in relation to net income (see also Bach et al. 2019). However, keeping a national carbon price "socially fair" has been one of the government's main arguments for starting with a price of only 10 euros per ton. This argument is misguided because a complete per capita reimbursement of a tax on carbon was not part of the government's proposal in 2019, even though this would have been more socially fair (see also Edenhofer et al. 2019; Bach et al. 2020). A tax revenue could be reimbursed per capita. Since the less affluent usually emit less carbon during their lifetime, they may even profit from this climate policy. It is possible that they may receive more money in return than the money they pay out in accordance with the increased tax on carbon-intensive activities. Due to this, a per capita reimbursement might even be progressive instead of regressive. Thus, a corresponding policy would have the potential to be fairer and consequently also more feasible. The reluctance to pay a higher price on carbon in a system without reimbursement can be seen as a reluctance to support an unfair policy. Moreover, the less affluent may even profit from a complete per capita reimbursement of a tax on carbon.

We should be confident that a fairer and more effective proposal would also be more acceptable to the majority of the voters if it is communicated well. Thus, I suggest to optimistically assume that fairness and political feasibility will prove to be linked in the long run. Furthermore, having an open public debate can be a major factor in increasing the justifiability and acceptance of certain policy decisions. Political participation contributes to a justified climate policy and thereby has the potential to enable its feasibility.

One could claim that the link between feasibility and justice is merely contingent: the reason that people are against a higher price on carbon may not be the injustice of the policy proposal per se but rather the costs they would have to pay if the policy was in fact implemented. Whether or not these costs are their fair share might not matter to them; they could just as well oppose such a policy due to the fact that they would have to bear any additional costs in the first place. Claims of this sort raise empirical questions about underlying motivations that I cannot answer in this chapter. Environmental psychology would be one of the relevant disciplines to inform us about these tendencies. Nevertheless, it seems quite obvious to me that people can *also* be against a certain policy *because* they judge it to be unjust. Although this might be especially likely if they themselves are the victims of this injustice, it seems plausible as well that people can be opposed to policies due to injustices that are directed toward others.

It is an empirical question whether or not the just alternative would be more feasible. Would a policy proposal that combines a higher price on carbon with a fair redistribution of the revenue be politically feasible? Would a majority of people support this proposal? We would have to find out. But finding out possible reasons for the reluctance of voters to support more ambitious climate policies seems to be an important step in realizing more climate justice. These reasons are probably not limited to motivational problems or a reluctance to do what is morally appropriate. So, it is worth giving it a try. If policy proposals are compatible with the relevant principles (such as the PPP and the APP), they may also be more convincing and thus politically more feasible.

III. FEASIBILITY VERSUS JUSTICE?

So far, I have argued that making a policy more just can promote its feasibility. It might be because of a policy's unfairness toward people with lower income that they do not support a higher tax on carbon. Alternative policies of carbon pricing could be fairer. And for that reason, they might also be more feasible. In this section, I will ask whether it could also be the other way around and I will explore how we should react to this option. How should we deal with the possibility that unfair policies could be more feasible?

Can justice considerations decrease the feasibility of corresponding policies? Although Patterson et al. (2018) generally suppose that social justice considerations have the potential to increase the political feasibility of climate policies instead of decreasing it, they admit that a focus on justice may also have unintended political consequences, such as being deployed *against* climate action. They do not decide these empirical questions, but they rightly point out that a social justice frame is likely to resonate differently among different groups with different agendas.

Regardless of these empirical questions, however, the fact that some unjust policies might be more feasible does not provide an overriding reason to support them. Good ends do not (always) justify evil means, and that is why being an effective means of reducing emissions is not all by itself a sufficient reason for promoting an unjust policy. Climate action and how it is communicated should be compatible with climate justice. From this perspective, the political consequences of pointing to reasons of justice are not "unintended." For example, if political parties point out that pricing carbon is unfair and if this leads to a resistance of the less affluent against setting a price on carbon, this can only be regarded as an unintended side-effect if a just climate policy is not what we are striving for. This resistance reveals a justified complaint that should be taken seriously at least by anyone who is concerned with making just policy decisions. Moreover, in taking it seriously, we can optimistically assume that fairer policy proposals would be confronted with less feasibility constraints.

In some cases, this is likely to be too optimistic. For instance, theories that demand complete responsibility for past emissions may be justified from a moral point of view, but to suppose that a corresponding policy would be equally feasible for that reason just seems unrealistic. However, the example of carbon pricing revealed that one should not give up optimism too quickly. Instead of prematurely accusing individual citizens of being reluctant to follow the demands of climate justice, it might be more productive to consider the possibility that their resistance to certain policy proposals might be the result of an injustice within these proposals.

Some contemporary theorists have a fundamentally negative attitude toward justice issues.[7] They argue that we should take power relations and given interests into account. I agree with this point, but I deny that moral and justice-theoretical questions should be excluded. And beyond that, I propose a fundamentally more optimistic picture: instead of focusing on the fact that a fair policy is often not feasible, we should take into consideration that sometimes a policy is not feasible because it is unfair.

IV. FEASIBILITY CONSTRAINTS AND IDEAL THEORY

The debate on ideal and non-ideal theory has for a large part revolved around Rawls's theorizing about justice. Thus much of the discussion focused on ideal and non-ideal theorizing about *justice* (Valentini 2012, 655). Despite this general discussion, climate ethics in particular also focuses on questions of justice and has also been accused of being merely ideal. Climate ethics is criticized for neglecting the feasibility of its proposals for a just climate policy.

In this section, I will argue against skeptics of ideal theory. They think that feasibility is an issue that speaks against ideal theory. The characterization of a theory as "ideal" is sometimes used as an accusation that the theory ignores important questions of feasibility (see, for example, Farrelly 2007). In contrast to setting feasibility and ideal theory apart, I will argue that we do not need less but more ideal theory in order to solve feasibility problems.

Are feasibility constraints largely neglected in climate ethics? Kowarsch and Edenhofer (2016) think so. They claim that ideal accounts of climate justice are an insufficient response to actual problems and complain that the climate ethics literature focuses on a few abstract ethical principles that are not embedded in a comprehensive evaluation of climate policy. In their opinion, the discussion of ethical principles has often been isolated from the decisions faced by policymakers, even though they have to consider various ethically relevant aspects of climate policy. The abstract principles of climate ethics are criticized for not being tailor-made to deal with these aspects. Kowarsch and Edenhofer claim that discussing only a few principles in climate ethics leaves policymakers uninformed when it comes to the difficult evaluation of climate policy pathways.

Posner and Weisbach (2010) have gone even further by arguing that the discussion of ideal climate ethics is not only disconnected from but also detrimental to real-world policy making. This is because insisting on achieving the demands of climate ethics is actually counterproductive and would doom any international climate change agreement.[8] More specifically, they refer to "emissions egalitarianism" (EE), a concept largely discussed in climate ethics that proposes a distribution of emissions that gives everyone an equal per capita share. Posner and Weisbach claim that this approach would be rejected by wealthy nations: "Per capita allocations would have the effect of redistributing hundreds of billions of dollars from wealthy nations, above all the United States, to developing nations. For this reason, insistence on per capita allocations would effectively doom any climate change agreement" (Posner and Weisbach 2010, 122). (Please see Weisbach, D. in this volume for more on this issue.)

Thus, it seems that some climate ethicists have argued that ideal theory is an unhelpful guide to actual practice in non-ideal circumstances. Moreover, it has even been accused of being detrimental to the realization of just policies. In what follows, I will defend ideal theory against this criticism by arguing that it is sometimes a valuable tool that can be used to overcome certain feasibility constraints. Not only does climate ethics suggest policies that are more just, but these policies have the potential to be more feasible due to their decreased level of injustice.

The distinction between ideal and non-ideal theory first entered contemporary political philosophy through the work of John Rawls. In *A Theory of Justice*, Rawls holds that "non-ideal theory" should be worked out after an ideal conception of justice has been chosen (Rawls [1971] 1999, 216). In *The Law of Peoples*, Rawls emphasizes the priority of ideal theory over non-ideal theory. In his view, non-ideal theory asks how the long-term goal of ideal theory might be achieved. It thereby depends upon the guidance of ideal theory:

> It looks for policies and courses of action that are morally permissible and politically possible as well as likely to be effective. So conceived, non-ideal theory presupposes that ideal theory is already on hand. For until the ideal is identified, at least in outline—and that is all we should expect—non-ideal theory lacks an

objective, an aim, by reference to which its queries can be answered. (Rawls 1999, 89–90)

Thus, according to Rawls, ideal theory provides the compass for non-ideal theory. He is also concerned with feasibility issues, as non-ideal theory looks for policies and courses of action that are politically possible as well as likely to be effective.[9]

According to this approach, ideal theory is an *end-state* theory, whereas non-ideal theory is a *transitional* theory.[10] Say we want to know how to reach a zero carbon society without compromising *other* forms of justice. For example, we have to add the perspective of domestic justice in order to evaluate the transition to a low carbon society. As we have seen, a very high tax on carbon in a system without reimbursement is an unfair policy.

In order to sustain this claim, I referred to ideal theory. I spelled out that placing greater burdens on the less affluent than on those who are more affluent is in conflict with plausible moral principles for the distribution of burdens in the context of climate change mitigation: the PPP and the APP. These principles are part of ideal theories of climate justice. However, not only concrete considerations within climate ethics can provide a justification for these principles but also more general considerations of justice. For example, Rawls claims that social and economic inequalities are to be arranged so that they are to the greatest benefit of the least advantaged.[11] This speaks in favor of the APP.

Contrary to the concerns raised by critics, I think that abstract principles of climate ethics can play a role in transitional contexts because they can provide the appropriate normative guidance for the transition toward a low carbon society. Thus, I deny the assertion that climate ethics is unable to guide climate policy.[12] (For more on the role of climate ethicists in policymaking, please see McBee, J. in this volume.) For example, abstract principles that are discussed in climate ethics can prove to be helpful in evaluating concrete policy proposals. As we have seen with carbon pricing in Germany, certain policy proposals are not supported by these principles. Instead, the feasibility of a policy proposal is strengthened insofar as it accords with these principles. Normative considerations are an important means by which to identify and analyze feasibility constraints. These constraints may not always be described as problems for an account of justice but rather as an expression of a form of injustice. Ideal theory can reveal that there are normative facts behind the descriptive facts. Moreover, these normative facts may even help explain the descriptive facts.

Non-ideal theory is often framed as a deviation from ideal justice. And just climate policies are often considered to be infeasible. For these reasons, some theorists have argued that considerations of justice should not be taken as relevant in the pursuit of a climate policy. However, this assumption should be taken with caution. Instead, I suggest a change in perspective: sometimes feasibility constraints are caused by a lack of justice.

V. CONCLUSION

Feasibility problems may be a function of the injustice of certain proposals. Climate ethicists should point to these kinds of injustices and not surrender to feasibility problems. Sometimes, climate ethicists may even contribute a solution to these problems by pointing to the relevant normative aspects of the situation, by suggesting fairer policies, and by explaining why they would be fairer.

For example, as shown earlier, it is possible for alternative policies of carbon pricing to be more socially just and thereby more feasible. Beyond this example, my conclusion could also be applied to further principles of climate justice and to further climate policies. There are other examples of policies that try to make climate action more feasible by addressing issues of injustice, for example the Green New Deal in the United States, which also addresses the perspective of those who work in the fossil fuel industry.

One should be aware of the fact that if some feasibility constraints diminish, others may arise. If non-ideal theory points to the difficulties of avoiding them (e.g., to the resistance of the rich to a just system of pricing carbon), it is an important correlate of ideal theory. The obstacles to a just climate policy are numerous, and besides morality, we should be acutely aware of the power dynamics at play. To be sure, lobbying groups also influence politics and we should not lose track of this fact.

However, instead of prematurely accusing individual citizens of being reluctant to follow the demands of climate justice, it might be more productive to consider the possibility that their resistance to certain policy proposals might be the result of an injustice within these proposals. In this chapter, I have suggested to "optimistically" assume that more just policy proposals might be confronted with less feasibility constraints. And it seems that for climate ethics, there is practically no alternative to this optimism. Climate ethics must assume that people are susceptible to relevant normative considerations.

This way to look at the world is not unreasonable, because holding onto this optimistic assumption can also contribute to its fulfillment. Pointing to the fact that even basic principles of climate justice are violated by certain policies may pave the way for more just and thereby for more feasible climate policies.

NOTES

1. See, for example, Roser and Seidel (2017) and Meyer (2018).

2. According to a different notion of feasibility, even an outcome that could be realized is considered infeasible if the moral costs of reaching it are too high (Räikkä 1998, 37). However, the question of whether a policy is infeasible and the question of whether its realization would be morally costly seem to be different questions (Southwood 2018). Thus, in this chapter, I will stick to the first approach. That is, by referring to a lack of feasibility, I intend to express that it is unlikely or even impossible that the desired outcome will be realized.

3. The Convention on International Civic Aviation exempts air fuels already loaded onto an aircraft from import taxes upon landing. And an EU-wide agreement on taxing aviation fuel may result in carriers filling their aircraft as full as possible whenever they land outside the EU in order to avoid paying tax. This might even increase the level of aviation emissions as the aircrafts will burn more fuel to carry the extra weight of a fully fueled tank. See (Seely 2019, 1).

4. Moreover, even if it is not possible to reach an agreement on the international level in a short amount of time, it is often possible to start with the national or European level. If certain measures on this level have economic costs, it is still a political decision whether or not these costs are accepted.

5. This view is almost universally shared in the philosophical literature on climate justice. See, for example, Caney (2012); Shue (2015); Roser and Seidel (2017).

6. For a summary of the process, see Wettengel, December 16, 2019.

7. Geuss (2008), for example, takes this perspective when he, without formulating an alternative, accuses theories of justice of lacking realism and being irrelevant.

8. For an overview of the relevant positions in climate ethics, see Roser and Seidel (2017). For an actual defense of Emissions Egalitarianism, see Torpman (2019).

9. From a similar perspective, Aldred (2016) argues that non-ideal theory is focused firmly on the path or transition to the ideal. Thereby, we should keep the ideal or long-term goal in mind. According to Aldred, emissions trading might be rejected by non-ideal theory because it obstructs our path towards the long-term goal of a low-carbon economy (Aldred 2016, 151).

10. This distinction can also be found in Valentini (2012).

11. Rawls (1971, 1999, 266).

12. In this respect, it serves the function of non-ideal theory as Stemplwoska understands the term. She asserts that non-ideal theory offers viable recommendations that are both achievable and desirable in the circumstances we are actually facing (Stemplowska 2008, 324, 330).

REFERENCES

Aldred, Jonathan. 2016. "Emissions Trading Schemes in a 'Non-Ideal' World." In *Climate Justice in a Non-Ideal World*, edited by Clare Heyward and Dominic Roser, 148–168. Oxford: Oxford University Press.

Bach, Stefan, Niklas Isaak, Claudia Kemfert, and Nicole Wägner. 2019. "Lenkung, Aufkommen, Verteilung: Wirkungen von CO2-Bepreisung und Rückvergütung des Klimapakets." *DIW aktuell 24.* https://www.diw.de/de/diw_01.c.683659.de/publikationen/diw_aktuell/2019_0024/mono.html.

Bach, Stefan, Niklas Isaak, Lea Kampfmann, Claudia Kemfert, and Nicole Wägner. 2020. "Nachbesserungen beim Klimapaket Richtig, aber immer noch Unzureichend: CO2-Preise Stärker Erhöhen und Klimaprämie Einführen." *DIW aktuell 27.* https://www.diw.de/documents/publikationen/73/diw_01.c.739525.de/diw_aktuell_27.pdf.

Caney, Simon. 2012. "Just Emissions." *Philosophy & Public Affairs* 40(4): 255–300.

Caney, Simon. 2016. "The Struggle for Climate Justice in a Non-Ideal World." *Midwest Studies in Philosophy* 40(1): 9–26.

Edenhofer, Ottmar, Christian Flachsland, Matthias Kalkuhl, Brigitte Knopf, and Michael Pahle. 2019. "Bewertung des Klimapakets und Nächste Schritte: CO2-Preis, Sozialer Ausgleich, Europa, Monitoring." *MCC Berlin.* https://www.mcc-berlin.net/fileadmin/data/B2.3_Publications/Working%20Paper/2019_MCC_Bewertung_des_Klimapakets_final.pdf.

Farrelly, Colin. 2007. "Justice in Ideal Theory: A Refutation." *Political Studies* 55(4): 844–864.

Frankfurter Allgemeine Zeitung. 2019. "Merkel Verteidigt Klimapaket: "Politik ist das, was Möglich ist"." Last Modified September 20, 2019. Accessed March 20, 2020. https://www.faz.net/aktuell/wirtschaft/nach-klima-strategie-einigung-merkel-verteidigt-klimapaket-16393990.html.

Gagnebin, Murielle, Patrick Graichen, and Thorsten Lenk. 2019. "Die Gelbwesten-Proteste: Eine (Fehler-)Analyse der französischen CO2-Preispolitik." *Agora Energiewende. Hintergrund.* https://www.agora-energiewende.de/veroeffentlichungen/die-gelbwesten-proteste/.

Geuss, Raymond. 2008. *Philosophy and Real Politics.* Princeton: Princeton University Press.

Gilabert, Pablo, and Holly Lawford-Smith. 2012. "Political Feasibility: A Conceptual Exploration." *Political Studies* 60(4): 809–825.

Jewell, Jessica, and Aleh Cherp. 2020. "On the Political Feasibility of Climate Change Mitigation Pathways: Is It Too Late to Keep Warming Below 1.5°C?" *WIREs Climate Change* 11(1): 1–12.

Kowarsch, Martin, and Ottmar Edenhofer. 2016. "Principles or Pathways? Improving the Contribution of Philosophical Ethics to Climate Policy." In *Climate Justice in a Non-Ideal World*, edited by Clare Heyward and Dominic Roser, 296–318. Oxford: Oxford University Press.

Lawford-Smith, Holly. 2013. "Understanding Political Feasibility." *Journal of Political Philosophy* 21(3): 243–259.

Meyer, Kirsten. 2018. *Was schulden wir künftigen Generationen? Herausforderung Zukunftsethik.* Stuttgart: Reclam.

Patterson, James J., Thomas Thaler, Matthew Hoffmann, Sara Hughes, Angela Oels, Eric Chu, Aysem Mert, Dave Huitema, Sarah Burch, and Andy Jordan. 2018. "Political Feasibility of 1.5°C Societal Transformations: The Role of Social Justice." *Current Opinion in Environmental Sustainability* 31: 1–9.

Posner, Eric A., and David Weisbach. 2010. *Climate Change Justice*. Princeton: Princeton University Press.
Räikkä, Juha. 1998. "The Feasibility Condition in Political Theory." *Journal of Political Philosophy* 6(1): 27–40.
Rawls, John. (1971) 1999. *A Theory of Justice* (Rev. ed.). Cambridge: Harvard University Press.
Rawls, John. 1999. *The Law of Peoples*. Cambridge: Harvard University Press.
Roser, Dominic, and Christian Seidel. 2017. *Climate Justice: An Introduction*. London: Routledge.
Seely, Anthony. 2019. *Taxing Aviation Fuel*. Briefing Paper, Number 523. London: House of Commons Library. http://researchbriefings.files.parliament.uk/documents/SN00523/SN00523.pdf.
Shue, Henry. 2015. "Historical Responsibility, Harm Prohibition, and Preservation Requirement: Core Practical Convergence on Climate Change." *Moral Philosophy and Politics* 2(1): 7–31.
Southwood, Nicholas. 2018. "The Feasibility Issue." *Philosophy Compass* 13(8): e12509.
Stemplowska, Zofia. 2008. "What's Ideal About Ideal Theory." *Social Theory and Practice* 34: 319–340.
Tank, Lukas. 2020. "The Unfair Burdens Argument Against Carbon Pricing." *Journal of Applied Philosophy* 37(4): 612–627.
Torpman, Olle. 2019. "The Case for Emissions Egalitarianism." *Ethical Theory and Moral Practice* 22(3): 749–762.
Valentini, Laura. 2012. "Ideal Vs. Non-Ideal Theory: A Conceptual Map." *Philosophy Compass* 7(9): 654–664.
Wettengel, Julian. 2019. "Germany's Carbon Pricing System for Transport and Buildings." Last Modified December 16, 2019. Accessed April 23, 2020. https://www.cleanenergywire.org/factsheets/germanys-planned-carbon-pricing-system-transport-and-buildings.

Chapter 6

The "Pathway Problem," Probabilistic Feasibility, and Non-Ideal Climate Justice

Jared Houston

Ongoing inaction on mitigation and adaptation has turned climate justice debates toward non-ideal theory, where the demand for action-guidance is foregrounded (Heyward and Roser 2016; Brandstedt 2019; Sayegh 2018). Some theorists have adopted an understanding of the problem of action-guidance toward climate justice as a "pathway problem." I argue that this theoretical framework's assumptions, including a reliance on probabilistic feasibility assessments, makes it inappropriate to guide action toward climate justice.

In this chapter, I begin by introducing non-ideal theory and the concept of feasibility. I then identify the "pathway problem" in the literature and explain its application to climate justice. Next, I argue that the pathway problem is an inappropriate theoretical framework for action-guidance toward climate justice. First, in a climate changing future, accurate probabilistic feasibility assessments become increasingly difficult, making the pathway problem impracticable. Second, the pathway problem inadequately models the political agencies relevant to climate justice. It fails to account for coalitions in climate politics, insufficiently models the influence of intelligent opposition, and makes assumptions that bias it toward the powerful. I then consider and respond to replies to this critique, including the claim that the pathway problem is "the best we can do." Finally, I suggest we can do better through *strategic climate planning*, an alternative framework in which probabilistic feasibility assessment is not integral to solving the problem of action-guidance. I conclude with a warning: strategic climate planning is already practiced by corporate and military agencies worthy of opposition.

I. NON-IDEAL THEORY, THE "PATHWAY PROBLEM," AND CLIMATE JUSTICE

Attention to non-ideal theory emerged as a methodological response to the prevalence of ideal theory in political philosophy. Critiques of ideal theory and the non-ideal methodological alternatives proposed are diverse (Valentini 2012; Farrelly 2007; Robeyns 2008; Hamlin and Stemplowska 2012; Sayegh 2018). I focus on the "action-guidance" critique, and proposals for theorizing action-guidance toward climate justice. Roughly, the demand for action-guidance states that the practical application of the normative outputs of political theories should effectively guide action in the real-world (Valentini 2009, esp. 341–343).

John Rawls's liberal-egalitarian theory has occupied a central role in debates about action-guidance. Rawlsian non-ideal theorists follow and expand upon Rawls's brief comments on the matter. Some argue that non-ideal theorizing should focus on *transitional plans* toward institutions satisfying an ideal of justice (Simmons 2010, sec. 3). Non-ideal theorists departing from Rawls hold that action-guidance is best developed without asserting the primacy of ideal theory (Sayegh 2018). Both camps agree, however, that feasibility is an essential consideration for action-guidance. Complementary to the normative desirability of proposed changes, feasibility considerations regard whether these changes are achievable in a specific political context. A wide literature has thus developed on the concept of feasibility. Theorists have distinguished different kinds of feasibility claims in politics, applicable to different actors, and (in)appropriate to different "stages" of theorizing (Hamlin 2017; Southwood 2018, sec. 2.2; Stemplowska 2016; Lawford-Smith 2012; Gilabert and Lawford-Smith 2012, 819–821). In this chapter, I focus on conceptions of feasibility that propose a strong connection between the demand for action-guidance and an assessment of the *probability* of achieving political outcomes.

The Pathway Problem and Probabilistic Feasibility

Toward fulfilling the action-guidance demand, non-ideal theory proposes transitional plans detailing the set of institutional arrangements traversed toward some (possibly ideal) end goal. A common analogy is climbing a mountain (Simmons 2010, 35; Gaus 2011, sec. 3.c; Hamlin 2017, figure 1). A mountain has a peak (the ideal), intermediate elevations (non-ideal institutional arrangements), and more or less passable terrain (feasibility). The mountaineer's task is to plan a route of ascent, factoring in the ultimate end goal, points of progression to intermediate elevations, and the likelihood of success in moving between those points. This analogy suggests that the aim of non-ideal action-guidance is to address this "pathway problem."

The pathway problem was suggested by Rawls,[1] but perhaps its most advanced articulation is offered by Pablo Gilabert and Holly Lawford-Smith (Gilabert 2017; Gilabert and Lawford-Smith 2012). There, the pathway problem comes last in a three-level theory of political feasibility covering (i) political principles, (ii) institutional implementations, and (iii) "strategies of political reform" or "trajectories of political change" (Gilabert and Lawford-Smith 2012, 820; Gilabert 2017, 109, respectively). Pathways proposed in (iii) are theoretically framed and evaluated using rational choice theory (also called decision theory):

> These kinds of considerations are familiar in rational choice theory, and we adapt them to the specific context of political judgment. (Gilabert and Lawford-Smith 2012, 822)

> We can adapt some guidelines from decision theory. For example, we can compute the expected value of alternative paths of transition (as well as the status quo) by considering their intrinsic desirability and probability of success and favor ones with [a] maximal score. (Gilabert 2017, 121)

Thus, candidate transitional plans consider both the desirability of institutional transitions and the likelihood (probabilistic feasibility) of their realization. The standard decision principle of rational choice—expected utility maximization[2]—is suggested to make "all-things-considered" judgments about the best pathway.[3]

If we adopt this framing of the pathway problem, then probabilistic feasibility is essential for guiding action. This is because principles like "expected utility maximization" call for accurate probabilities as inputs. Thus, non-ideal theorists appealing to this framework must draw on evidence to develop accurate probabilistic feasibility assessments of the likely success of each stage of their transitional plans. Any uncertainty in these probabilistic feasibility assessments affects the "all-things-considered" assessment of pathways of institutional change.

Probabilistic Feasibility and the Pathway toward Climate Justice

So understood, the pathway problem is claimed to apply to political judgments generally, including considerations of climate justice.[4] Climate justice is concerned with three primary obligations: mitigation, adaptation, and compensation (see for example Caney 2016, 22; Moellendorf 2015). As climate change is driven by anthropogenic greenhouse gas emissions, their rapid reduction is demanded by mitigation.[5] But historical failures to reduce emissions mean that climate changes are ongoing and impending. Thus, adaptation is also required to reduce the negative impacts of these changes. Compensation is required for the remainder of negative impacts (i.e., loss and damage) that cannot be sufficiently minimized with adaptation measures. The pathway problem for climate justice would thus assess alternative transitional policy approaches toward fulfilling the duties of mitigation, adaptation, and compensation.

Non-ideal theorists addressing the pathway problem of climate justice reassert the integral role of probabilistic feasibility. Johnathan Aldred applies Rawlsian non-ideal transitional theory to climate justice and argues that the proper focus of climate policy should be the long-term transition to a low-carbon economy. Aldred argues that emission trading schemes are inferior to the direct regulation of carbon emissions because the former are *unlikely* to achieve a low-carbon economy in the long-term (2016, sec. 7.2-3). Simon Caney outlines a methodological framework to respond to noncompliance with climate justice that combines normative analysis with an assessment of the *chances* of a proposal being successfully implemented (Caney 2016, 35–37). Alexandre Gajevic Sayegh develops an "integrative" non-ideal theory and applies it to the case of climate change. Its third theoretical level analyzes "political processes," including the *likelihood* of achieving institutional change, such as mitigation policy (2018, 412–413).

II. THE PATHWAY PROBLEM IN A CLIMATE-CHANGING FUTURE: A CRITIQUE

As modeled by probabilistic rational choice theory, the pathway problem offers decision-support to agents enjoying detailed knowledge of both the outcomes and probabilities of actions. The agents are unafflicted by real-world complexities such as inadequate or poor information, epistemic limitations, external surprises, and intelligent opposition. Finally, they act from a position of relative power and control. Each of these assumptions belies the real-world complexities of pursuing justice in a climate-changing future. The pathway problem, as theorized through probabilistic rational choice theory, is thus an inappropriate action-guidance framework for climate justice.

Uncertainty and Probabilistic Feasibility Assessments in a Climate-Changing Future

A climate-changing future is fraught with considerable *uncertainty* that undermines the prospect of accurate probabilistic feasibility assessments demanded by the pathway problem.[6] An example will facilitate my argument. Consider three phases in the pursuit of a "green new deal" in the United States. First, a national campaign group is formed that lobbies several major cities to pass municipal "green new deal" legislation. It is hoped this builds pressure at the state level, where in the second phase, the campaign pushes for broader legislation. Finally, with multiple states passing aggressive emissions reduction regulations, the growing national campaign makes a final push for federal legislation to regulate emissions nationally, transition to a green economy, and adapt to impending climate changes. Modeled through the probabilistic rational choice framing of the pathway problem, each of the three phases of this plan would have its desirability evaluated, as well as the probabilistic feasibility of transitions between them. The results would be compared to alternative pathways, toward the goal of maximizing expected normative outcomes. Uncertainties begin with the sense that it will be harder to accurately assess the probabilistic feasibility of the transition to the third phase than to the first. We not only have less information about the far-off future to inform our assessment than we do about the near-term, but the institutional maneuvers in phase three are broader in scope, more complex, and involve more political actors.[7] Given the extent of the institutional changes needed to address the climate crisis, and the time needed to implement them, these considerations seem true for any morally ambitious climate justice pathway.

These initial difficulties regarding the accuracy of probabilistic feasibility assessments are compounded when we consider the pending biophysical, social, and economic impacts of climate change. The Intergovernmental Panel on Climate Change's (IPCC) 5th assessment report offers multiple discussions of abiding uncertainties, and the treatment thereof, in the modeling of the potential biophysical impacts of climate change over the twenty-first century (Working Group I 2013, sec. 1.4, 12.2.2; Working Group III 2014; Mastrandrea, M.D. et al. 2010). I will expand on two: model uncertainty and emissions scenario uncertainty (Working Group I 2013, sec. 1.4.2). The complex dynamics of the climate as a biophysical system make it difficult to model. Choices of climate model structure and parameters are not determinate, leading to a plurality of valid models and results (Betz 2009). To project climate changes and impacts, models take emissions scenarios as inputs, which express uncertainty about future human behavior regarding mitigation and energy policy (Working Group I 2013, sec. 12.3). These uncertainties are expressed through several "representative [GHG] concentration pathways" (RCPs) over the twenty-first century. These two sources of uncertainty combine in model ensembles to produce uncertainty in key climate change indicators, such as global mean surface temperature (GMST) increases, which are projected to range between +0.3 and +4.8°C by 2100 (see figure 6.1, adapted from Working Group I 2013, 1054, fig. 12.5).

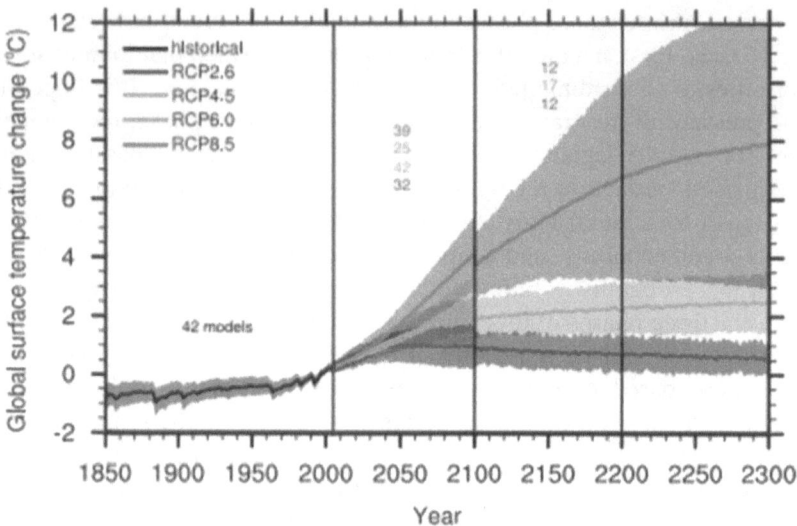

Figure 6.1 Global annual mean surface air temperature anomalies (relative to 1986–2005) from CMIP5 concentration-driven experiments.

The biophysical impacts of +0.3°C GMST differ greatly from +4.8°C, as do the correlative social and economic disruptions. Given the interdependencies between biophysical, social, and economic systems, and political institutions, uncertainties in the range of impacts of climate change introduce further difficulties in the accurate assessment of the political feasibility of institutional transitions.[8] Returning to the "green new deal" example, the probabilistic feasibility of moving from phase two to phase three seems higher if actual GMST tracks the higher-end of the range, with the correlative negative impacts (e.g., storms, floods, droughts, heatwaves) creating a sense of federal urgency to legislate on the problem. If the impacts are more moderate, however, then this sense of urgency is lacking, and the probabilistic feasibility of a move from phase two to three seems lower. Uncertainty about climate impacts thus translates into uncertainties in probabilistic feasibility assessments, which become increasingly difficult, especially over the long-term. As the accuracy of the probabilities it requires as inputs diminishes, the probabilistic rational choice framing of the pathway problem becomes increasingly inoperable.

To the previously stated difficulties regarding (i) the need for extensive, multi-stage institutional change over the long-term, and (ii) uncertainties in the response of the climate system to historical and ongoing GHG emissions, I would add (iii) the abiding unpredictability of politics. Consider events like the fall of the Soviet Union, the 2008 financial crises, or the recent

COVID-19 pandemic. Biased by hindsight, it is easy to underestimate how unexpected and disruptive these events were to ongoing transitional political projects. The vulnerability of political systems to unpredictable and disruptive events that radically change feasibility assessments does not look set to diminish in a climate-changing future. Indeed, climate change may amplify political volatility, for example through its role as a polarizing issue that erodes confidence and interest in solving problems democratically, or through increasing the threat of violent conflict (Hsiang et al. 2013).

With these three difficulties combined, I am skeptical that we can accurately assess the probabilistic feasibility of multi-state, long-term transitional pathways in a climate-changing future as the rational choice model demands.

Inadequate Modeling of the Agents Relevant to Climate Justice

Action-guidance for climate justice must also take proper account of the multiple competing agents involved in the climate crisis. Consider a tripartite characterization of these agencies: deciding, opposing, and dominated. Deciding agents are sensitive to the moral demands of climate justice and enjoy the power to enact political change. Citizens, states, and civil society organizations working toward climate justice, and typically among the global affluent, would fall under this category. Opposing agents enjoy powers similar to Deciding agents but are either insensitive or hostile to the demands of climate justice. Some fossil-fuel corporations, morally callous politicians beholden to the former's sway, and climate-denying citizens among the global affluent, would fall under this category. Dominated agents may be sensitive to the demands of climate justice, but lack the power enjoyed by the two former agencies. The global poor fall under this category. All these agents are inadequately theorized in the pathway problem, and probabilistic feasibility assessments are implicated in this inadequacy.

Deciding Agents

Probabilistic rational choice theory offers action-guidance to a single agent.[9] But the social and political actors working toward climate justice in a non-ideal world are not single agents. Instead, they are often fractious coalitions of many groups.[10] Ignoring the internal diversity of these groups means ignoring issues relevant to non-ideal action-guidance.

The different groups comprising deciding agents will likely have different rankings of the desirability of proposed institutional transitions, and different assessments of their probabilistic feasibility. Decisions regarding the best transitional pathways will thus exacerbate tensions. These tensions cannot be resolved by demanding a unified normative theory, or by appealing to "the best" interpretation of evidence to settle disagreements about feasibility, as this rejects pluralism (both normative[11] and empirical[12]) as a basic feature of coalitions. Rejecting pluralism negatively impacts the integrity and efficacy of coalitions given that alienated subgroups will exit the coalition if their frustrations regarding its lack of accommodation of their views become intense. Thus, there must be an institutional means of resolving disagreement that respects pluralism. Non-ideal theory might help here, but in doing so, must depart from the "single agent" model of the pathway problem.[13]

Opposing Agents

A more striking oversight in the probabilistic rational choice framing of the pathway problem is the inadequate modeling of opposing agents. Probabilistic rational choice theory models the relation between a single deciding agent's subjective beliefs and preferences, external states of affairs, and outcomes of the agent's actions in these states (see e.g., Peterson 2009, ch. 2). All features of the decision context must be reducible to one of these elements. For example, say I am deciding whether to bring my umbrella with me to work today or not (action). I know it will either rain or not (external states of affairs). I form the subjective belief that there is a 90 percent chance of rain. Given that I prefer to be dry rather than wet more than I don't want to be burdened with my umbrella (subjective preferences over outcomes), and that rain is likely, I maximize my expected outcome by bringing my umbrella. In this example, there are no opposing agents using their intelligence to thwart my plans (say by misinforming me about the chance of rain). In pursuing climate justice in a non-ideal world, however, there are such agents. I argue that the impact of these agents is inadequately modeled in the single-agent probabilistic rational choice framework.

Opposing agents are often modeled in the probabilistic rational choice framing of the pathway problem by representing their impact as negative contributions to the probabilistic feasibility of institutional transitions. For example, in the absence of climate skepticism's influence over the US Congress, aggressive mitigation policies may have been probabilistically feasible (e.g., ~60%), but the prevalence of such skepticism decreases that probability (to e.g., ~30%).[14] Reductionist modeling through probability decrements is inadequate as it ignores the intelligent capacities of opposing agents (e.g., surveillance, deception, strategic flexibility) and their willingness to use these capacities to thwart opponent's plans. As George Tsebelis warned long ago, the overconfident and fallacious modeling of intelligent actors through probabilities is common in political analysis (1989).

Thus, the powerful real-world agents successfully opposing action on mitigation, adaptation, and compensation are not given a prominent role in the probabilistic rational choice framing of the pathway problem. This oversight is not only naïve but dangerous. To be resisted, those opposed to climate justice *must* be named: carbon capitalists, political actors beholden to their economic power, and propagandists pushing climate denialism. When the presence of opposing agents is obscured, and their impacts inadequately modeled, we cannot provide useful action-guidance in the face of their powerful opposition. Rather than reducing the influence of the opponents of climate justice to decrements in probabilities, non-ideal action-guidance must name and engage them as a serious threat. To do so, it must move beyond probabilistic rational choice framings of the pathway problem.

Dominated Agents

The probabilistic rational choice framing of the pathway problem alleges to provide general council to political agents seeking action-guidance. But it makes two assumptions about agents that bias it toward the capacities and abilities of powerful agents. First, agents are assumed to enjoy significant institutional capacities to effectively integrate probabilistic information into political and policy judgments. Second, agents are assumed to only be minimally vulnerable to disruptions of their plans. These assumptions make the probabilistic rational choice framing of the pathway problem biased against the global poor. Because of their position in global economic and political systems dominated by the affluent, the global poor are disadvantaged in their institutional capacities, and vulnerable to political and climactic volatility.

In climate treaty negotiations, least developed countries (LDCs) are disadvantaged in their diplomatic capacities and have struggled to have their interests heard (Abeysinghe and Huq 2016, sec. 9.3.3). After seeing partial success in their treaty demands,[15] the United States—a leading greenhouse gas emitter—announced it would withdraw from the Paris Accord (Friedman 2019). If LDCs had plotted a probabilistically feasible pathway to transition institutions toward climate justice assuming US fidelity to the treaty, this revelation would significantly disrupt that pathway. The probabilistic rational choice framing of the pathway problem fails to account for the global poor's vulnerability to disruption of their plans brought on by the erratic will of powerful agents.

Developing countries also face difficulties in exercising the institutional capacities needed to properly integrate probabilistic weather forecasts into adaptation planning (Agrawala et al. 2001, 466–467). Local communities are more vulnerable to the impacts of error in these forecasts: "What might be a plausible aberration in terms of frequent probabilities to a climate scientist might, in fact, cost some users their entire livelihoods" (Broad and Agrawala 2000, 1694). In disseminating probabilistic forecasts, scientists worry that their role in adaptation planning favors powerful commercial institutions in developing countries, who enjoy superior abilities to integrate probabilistic information into planning and are less vulnerable to errors.[16]

The lack of institutional capacity to integrate probabilistic information into political judgment, and high vulnerability to error, impact not only adaptation policy specifically but the value of the probabilistic rational choice framing of the pathway problem for LDCs more generally. Probabilistic feasibility assessments in the pathway problem are often complex syntheses of more acute probabilistic information. For example, to assess the probabilistic feasibility of a national freshwater management plan, LDCs might need to synthesize probabilistic information about the relation between water stress, violent conflict, and institutional stability, placing demands on their (limited) institutional capacities. And when pathways are pursued, LDCs remain especially vulnerable to errors and disruptions, given the threats of political and economic volatility these pose to sometimes fragile institutions.

Perhaps aid should be offered to the global poor to enhance their institutional capacities to integrate probabilistic information and reduce their vulnerabilities. Indeed, these efforts are underway through climate financing for adaptation; the problem, however, is two-fold. First, as of 2016, financing promises remained unfulfilled (Abeysinghe and Huq 2016, 196–197). Second, and more importantly, the suggestion that the global poor should rely on aid from the affluent is suspect given the moral relationship of domination. The institutional damage of colonialism, cold-war militarism, and forced neo-liberalization have systemically disadvantaged the global poor while benefiting the affluent. Global justice advocates decry the huge burden of debt service placed on the global poor by financial institutions dominated by the affluent (Ndikumana and Boyce 2011). It is thus morally scandalous that the affluent offer more than half of adaptation financing as loans, not grants (Abeysinghe and Huq 2016, Table 9.1). The global poor may therefore instead consider shifting to a non-probabilistic action-guidance alternative, appropriate to their circumstances, that does not increase their vulnerability to predation by the affluent.

The probabilistic rational choice framing of the pathway problem is thus an inappropriate action-guidance model for climate justice. It requires accurate probabilistic feasibility assessments that become increasingly difficult due to climate change. It inadequately models the real-world diversity and complexity of the political agents pursuing and opposing climate justice and is insensitive to entrenched inequalities in power. Notably, the overconfident application of probabilistic rational choice theory was warned against by its progenitors. As Ken Binmore explains, Leonard Savage made clear the model was only appropriate to "small worlds":

> According to Savage, a small world is one within which it is always possible to "look before you leap". [The deciding agent] can then take account *in advance* of the impact that all conceivable future pieces of information might have on the underlying model that determines her subjective beliefs. Any mistakes built into her original model that might be revealed in the future will then *already* have been corrected, so that no possibility remains of any unpleasant surprises. . . . In a large world, the possibility of unpleasant surprises that reveals some consideration overlooked in [the deciding agent's] original model can't be discounted. (2008, 117)

The pathway problem for climate justice is not a "small world" problem. It involves extensive, long-term institutional transitions in a complex and conflictual political context made more so by impending climate changes. Considering the aforementioned warning about the misapplication of probabilistic rational choice theory to "large worlds," Gilabert and Lawford-Smith's comments on the potential for error in their appeal to that theory are revealing:

> we may turn out to be wrong about what was, all things considered, the best choice, when more information comes to light. *But our error is permissible; it is the best we can do.* (2012, 823, my emphasis)

I disagree. Errors are not permissible if upon investigation they can be attributed to flawed assumptions in the action-guidance model and resolved by shifting to an alternative. Overconfidence in the probabilistic rational choice framing of the pathway problem threatens the efficacy of our pursuit of climate justice. It advocates contextually inappropriate planning tools that underestimate uncertainty, and overlook the abiding potential for errors, surprises, and unintended consequences. In doing so, it makes plans more vulnerable to disruption, and sacrifices preparedness and flexibility to maximize expected outcomes.

III. DEFENDING THE PATHWAY

I now consider two defences against my critique of the probabilistic rational choice framing of the pathway problem drawn from recent work by Dominic Roser (2017).[17] The first defends against the critique that accurate probabilistic feasibility assessments are unavailable, by permitting the use of low epistemic credential probabilities. The second defence claims that there is no adequate alternative to probabilistic rational choice, and thus despite its flaws, that framing of the pathway problem remains our best action-guidance theory.

Probabilities Need Not Be Accurate

Roser argues that the distinction between situations of risk where probabilities are available, and uncertainty where they are not, is fallacious.[18] He argues that probabilities are always available for action-guidance because they can and should be used regardless of their epistemic credentials:

> Low epistemic credentials are better than no credentials. If probabilities with high epistemic credentials are unavailable, then probabilities with low credentials are second-best. (2017, 14)

Thus, the critique made above—namely, that accurate probabilistic feasibility assessments are not available for the probabilistic rational choice framing of the pathway problem—can be avoided by denying that accuracy is required.

But in relaxing the epistemic credentials needed for probabilistic feasibility, we threaten the action-guiding effectiveness of the probabilistic rational choice framing of the pathway problem by making it vulnerable to errors of judgment stemming from probabilistic feasibility assessments based on low-quantity and/or poor-quality information.[19] Roser seems to admit this problem, and offers a reply:

> *ex post* we often realize that what we justifiably believed to be evidence actually had a net effect of misleading us rather than directing us towards the truth. But since decision-making principles instruct us from an *ex ante* perspective this is not relevant for the issue at hand. (2017, 14)

Yet, we must be permitted to assess the effectiveness of action-guidance frameworks by integrating critical observations about past implementation into future application. Yesterday's ex post realizations that our probabilistic feasibility assessment was in error are today's ex ante considerations about how to improve action-guidance to mitigate future errors. Like all normative guidance, decision principles are prospective in their outputs, but this does not imply that we should blind ourselves to the history of their implementation in (re)designing the process that generates those outputs.

No Adequate Alternative

Probabilistic rational choice theory is sometimes thought to set the boundaries of rationality itself, any departure from it being irrational (Clarke 2006,

41). This makes for a ready defence against critique: probabilistic rational choice "is the best we can do" (Gilabert and Lawford-Smith 2012, 823). This response can admit the critiques made previously and maintain that without an adequate alternative action-guidance framework, they are flaws we must live with.

To show a lack of adequate alternatives, Roser argues that popular rational decision principles alleging to operate without probabilities are flawed. Targeting the precautionary (maximin) principle, Roser first argues that precautionary theorists must distinguish realistic from unrealistic possibilities (see Gardiner 2006, sec. 9.B). In doing so, they make tacit appeal to probabilities, and thus precautionary reasoning is not an alternative to probabilistic methods.

> Judging a consequence to have a certain minimal level of evidential support, however, precisely means judging it to have a certain minimal probability. Thus, realism judgements are nothing but probability judgements. "Realistic" is just another word for "not highly improbable." (Roser 2017, 16)

This argument appears to suggest that evidential support claims *must* be expressed through a Bayesian logic. But there are possibility logics that can express these claims (Dubois and Prade 2015). One such logic, possibility theory, offers two measures regarding the occurrence of an event, its possibility (i.e., consistency with general knowledge of process; degree of unsurprise regarding occurrence) and its necessity (i.e., certainty; degree of surprise regarding non-occurrence). Probability theory uses only a single measure of uncertainty (Dubois and Prade 2007). To illustrate the contrast, consider the claim "GMST increase will exceed 4°C by 2100." Suppose probability theory measures the likelihood of this occurrence as 5 percent, (e.g., by assessing the relatively few climate model results that detail such heating along with the prior probability of the emission pathways they assume). Possibility theory assesses this claim through two measures instead of one, and with appeal to different evidence in each case. For example, GMST increase above 4°C by 2100 might be assessed as 80 percent possible (we know such temperatures are consistent with our general knowledge of the climate system, in this sense it would be mostly unsurprising if it occurred) and 20 percent certain (climate models don't uniquely imply this result, it would be minimally surprising if it did not occur). Probability theory, by assessing all evidence on a single scale, gives only indirect attention to controversies regarding what is realistically possible. In formalizing its analysis with two uncertainty measures, possibility theory attends directly to the distinct evidential analysis bearing on the important task of delineating realistic from unrealistic possibilities. With possibility theory as a well-developed alternative treatment of uncertainty, no appeal to probabilities is necessary in marking the boundary of realistic possibility. The forced translation of non-probabilistic evidential support claims into probabilistic forms by Bayesians obscures alternative treatments of uncertainty threatening the "no adequate alternative" claim.

In a second argument, Roser claims that since probabilities are (tacitly) used to distinguish realistic from unrealistic possibilities, it is arbitrary to deny their further use for assessing the likelihood of scenarios.

> If our evidence is such as to allow for a judgement about the realistic range of consequences, this same evidence surely allows for at least some probability judgements within and beyond that range. (2017, 16)

I will contest this claim by applying it to uncertainties in the projection of GMST increase discussed earlier. The IPCC claims it is likely (66%) that GMST increase will range between 0.3 and 4.8°C by 2100 (Working Group I 2013, 1055, Table 12.2). Thus, we conclude that global cooling is unrealistic, as is heating over 5°C. The evidential standard at work here is the distribution of climate model results.[20] According to Roser, appealing to these "epistemic credentials" to identify the boundaries of realistic GMST increase allows one to use the same evidential standards to make probability judgments within that range. Thus, we should be able to assign a probability to a GMST increase of 1 to 2°C, for example.

Problems for this claim begin with the fact that the IPCC's range of climate model results are generated using representative concentration pathways (RCPs) as inputs (Working Groups I, II, and III 2014, 57, Box 2.2). These pathways model uncertainty about future atmospheric GHG concentrations, and correlative radiative forcing values, through four different scenarios ranging from mitigation (RCP 2.6), to very high emissions (RCP 8.5), with two intermediate scenarios (RCP 4.5 and 6.0). If Roser is correct, we should be able to determine the probability of a GMST increase of 1 to 2°C by assessing the distribution of model results in that range; that is, invoking the same epistemic credentials used to probabilistically bound the realistic range of 0.3 to 4.8°C (see figure 6.2, adapted from Working Group I 2013, 1058, fig. 12.8).

But there is a difference, epistemically, between assessing the full range of model results across all RCPs and assessing a specific interval within those results. The difference, for the IPCC, is that while we can be confident that the RCPs and model ensemble together capture our uncertainties about atmospheric GHG concentrations, the same confidence does not carry to any individual scenario or specific interval of results, for two reasons. The first is because confidence in a fine-grained estimate of the probability of GMST increase between 1 to 2°C is crucially conditioned by the background

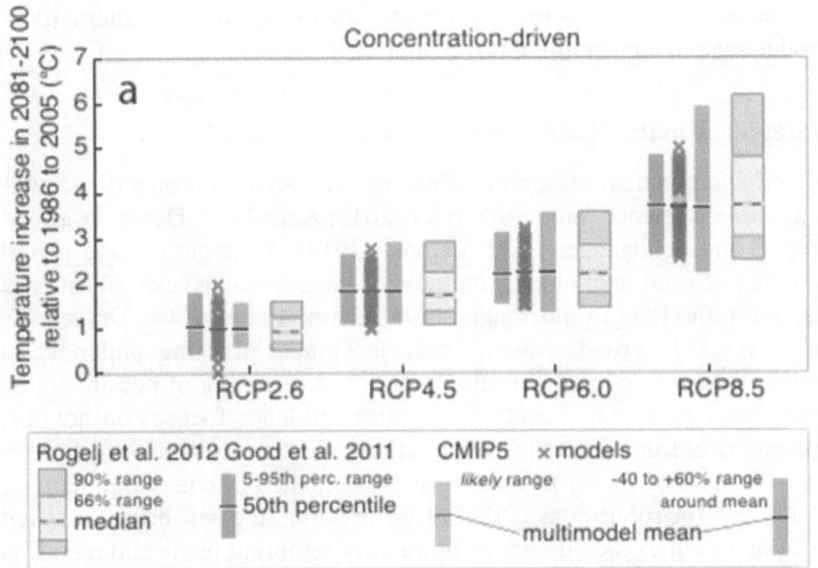

Figure 6.2 Uncertainty estimates for global mean temperature change in 2081-2100 with respect to 1986–2005.

probability that the RCPs used in modeling those results actually come about. It is precisely the difficulty in answering that key question (again driven by uncertainty about human behavior regarding emissions) that motivates the methodology that explicitly states probabilities are not assigned to RCPs.[21] Second, probability statements cannot be made about any interval of warming by assessing the distribution of model results therein without assessing bias and interdependence between models producing those results (Working Group III 2014, sec. 2.5.7.2). If the models share biases or dependencies, then results are not truly independent, which undermines the fine-grained probabilistic analysis. Thus, contrary to what Roser claims, warnings against finer-grained probabilistic judgments within the realistic range are not arbitrary. The RCP methodology and concerns about model bias and interdependence, give us non-arbitrary reasons to be confident in the realistic range of projected GMST increases (0.3 to 4.8°C) while withholding inferences toward the likelihood of specific intervals in the range (e.g., 1 to 2°C).

IV. DEPARTING THE PATHWAY

Since alternatives to the probabilistic rational choice framework for action-guidance have scarcely been surveyed, it's premature to claim none of them could be adequate. In this section, I contribute to the survey of alternatives by introducing and advocating a novel approach.[22]

Strategic Climate Planning

Possibilistic practical reasoning offers an alternative to probabilistic analytics in climate science and policy (Oels 2013; Betz 2010; Dessai et al. 2009; Lempert and Schlesinger 2000; Lempert 2013). Strategic climate planning combines several techniques including scenario analyses, robust strategies, and reflexivity (Schoemaker 1995; Lempert et al. 2006; Dryzek 2016, respectively). In broad strokes, strategic climate planning undertakes the empirically informed and narratively enriched visioning of possible climate futures through scenario analysis.[23] Action-guidance focuses on how organizations could develop robust institutions that fare well across these scenarios, instead of trying to maximize performance in one. Complementing the demand for robustness is that of reflexivity, "the self-critical capacity" to respond to unexpected change by rapidly rerouting plans and reconfiguring institutions (Dryzek 2016, 6). These capacities are crucial because a climate changing future will present a strategic context fraught with uncertainty, unpredictability, volatility, and rivalry. In such conditions, charting a probabilistically feasible pathway of institutional change is not practicable given the abiding threat of error, disruption, surprise, and unintended consequences. Climate justice demands a framework for action-guidance that, instead of idealizing these real-world complexities away, integrates their dynamics into planning and shifts action-guidance toward robustness and reflexivity.

Strategic climate planning offers improvements over the probabilistic rational choice framing of the pathway problem. Difficulties in accurately assessing probabilistic feasibility do not stall action-guidance, which can instead be delivered using relatively accessible possibilistic feasibility assessments (discussed later). The fractious politics of coalitions (see "Deciding Agents") are navigated by participatory planning that accommodates normative and empirical pluralism by visioning and preparing for multiple futures. Strategic climate planning takes seriously the power of those who oppose climate justice (see "Opposing Agents") and demands we study their strategies to effectively resist them. It accounts for the disruptive threats they pose by planning for contingencies in our struggles against them. Finally, strategic climate planning takes seriously power inequalities in the geopolitics of the climate crisis (see "Dominated Agents"). It does not strain the institutional capacities of the global poor by requiring probabilities for planning, but instead offers action-guidance appealing to ranges of possible climate changes. Furthermore, it accounts for the abiding disruptive threats that dominating agents pose, aiming to reduce vulnerability to interference through robustness and reflexivity.

Kyle Powys Whyte (et al.) presents a case study exemplary of strategic climate planning (2014). Their project aimed to support adaptation planning in Tribal communities in the Great Lakes region of the United States. The team designed a scenario planning workshop that presented "plotlines" including "challenging," "confounding," and "positive and transformational" possible futures (Whyte et al. 2014, 5). Instead of requiring the community to engage in probabilistic planning, participants were offered a more accessible, inclusive, and culturally appropriate way of thinking about the impacts of climate change through empirically informed creative narration.[24] After assessing and modifying the "plotlines," workshop participants used them as the descriptive bases to reflect on strategies to build preparedness across this range of possible climate futures.[25]

Strategic climate planning combines high normative ambitions toward climate justice with a focus on adaptability and preparedness for climactic and political volatility. Its action-guidance framework is accessible and relevant to a broad range of organizational actors. Perhaps most importantly, it is grounded in practice (Rickards et al. 2014; Moore et al. 2013). Instead of prescribing engagement with unfamiliar and idealistic rational choice models, we should consider seeking philosophical insights from, and offering refinements to, action-guidance frameworks already adopted by organizations pursuing climate justice.

What role remains for feasibility in strategic climate planning? Probabilistic feasibility assessments, as applied to complex systems over the long-term, are unreliable and unnecessary for action-guidance. While probabilistic feasibility may still have relevance to a limited set of action-guidance problems, it is not integral to strategic climate planning. Possibilistic feasibility assessment, however, is integral. To illustrate the contrast, suppose our organization has decided eco-socialism is the best solution to the climate crisis. The probabilistic rational choice framing of the pathway problem would recommend that we dedicate time and resources to accurately assessing the probabilistic feasibility of alternative institutional pathways as to maximize expected normative outcomes. Strategic climate planning would instead recommend that time and resources be shifted to mapping the realistically possible range of future scenarios we might need to be prepared to flexibly adapt our eco-socialist program to, due to the complex interaction of our own independent actions, our opponent's maneuvers, and unpredictable external circumstances. Such mapping is supported by the evidential assessment of realistic possibilities, involving (for example) the interpretation of climate model results, not as fine-grained probabilistic predictions of future climate changes, but as offering a broad understanding of the range of possible climate futures we might face (Betz 2010; Katzav 2014). Treating the evidence offered by climate models as useful for mapping realistic physical possibilities in the climate system, instead of assigning probabilities, is an example of the shift to possibilistic feasibility assessment. Recent research suggests further methods. Internal consistency analysis can help construct highly consistent scenarios to be prioritized for planning (Lloyd and Schweizer 2014). Non-probabilistic concepts such as "production possibility frontiers" can help assess organizational capacity against resource constraints in bringing about desired outcomes, refining the mapping of political possibilities (Wiens 2015).

Theorists have claimed that possibilistic feasibility assessment is relatively uninteresting, based on a conception that restricts it to simple and uncontroversial logical, metaphysical, or nomological "hard" constraints (Lawford-Smith 2013, 254). But the research cited earlier challenges this restricted characterization and suggests a broader conception of possibilistic feasibility worth developing. In shifting to strategic climate planning, possibilistic feasibility becomes more interesting because evidential matters previously theorized as exclusive to probabilistic analysis are open for possibilistic analysis.

V. CONCLUSION: A WARNING AND AN OPPORTUNITY

I have argued that the probabilistic rational choice framing of the pathway problem is inappropriate for action-guidance in a climate-changing future. Compounding uncertainties frustrate the accurate probabilistic assessment of the feasibility of institutional transitions. The diversity and complexity of political actors pursuing and opposing climate justice is inadequately modeled by rational choice theory. When approached with these flaws, advocates of the probabilistic rational choice framing of the pathway problem might reply: "there is no adequate alternative." I argued against that claim, with specific attention to strategic climate planning as an alternative with practical grounding and theoretical appeal.

I have stressed that in pursuing climate justice, we must be wary of our opponents. Heeding my own advice, I therefore close with a warning. Strategic planning and scenario analysis, as modes of possibilistic practical reasoning, originated and were refined in the high-uncertainty, high-rivalry contexts of commercial and military planning (Chermack 2005; Brown 1968). Private consultants, military agencies, and independent think-tanks are far ahead in strategically planning for climate change (Schwartz and Randall 2004; Klare 2019; Campbell et al. 2007; Spratt and Dunlop 2019). Unsurprisingly, they prioritize the analysis of climate change as a threat to national security. Unopposed, their narrow assessments will push climate policy toward "militarized adaptation" and a "politics of the armed lifeboat" (Parenti 2011, 11, 20). Political philosophers thus have the opportunity to critically engage national-level strategic climate planning with neglected concerns for "equity and justice" (Oels 2013, 27). Instead of rejecting possibilistic practical reasoning as "irrational," we could recover and redeploy it against those insensitive to the demands of climate justice (Houston 2020). Thus, there is also an urgent practical reason for further research into possibilistic action-guidance alternatives: to understand and counter our opposition.

NOTES

1. "Nonideal theory asks how this long-term goal might be achieved, or worked toward, usually in gradual steps. It looks for courses of action that are morally permissible and politically possible as well as likely to be effective." (Rawls 1999, 89)

2. Roughly, expected utility maximization instructs the rational chooser to select the alternative with the highest product of net utility (i.e., desirability) and probability.

3. "We have reason to engage in prospective choice seeking to maximize expectable normative value" (Gilabert 2017, 121). Though earlier, Gilabert expresses doubts that an "algorithmic decision procedure can be identified" (2017, 121).

4. Gilabert and Lawford-Smith claim the elements of their framework "form a set of considerations that should orient an all-things-considered judgment with respect to any path of political action" (2012, 822).

5. The distribution of the benefits and burdens of mitigation (and other duties) have been analyzed with reference to established theories of distributive justice. Debates in that field thus re-emerge as applied to the problem of climate change (see e.g., Shue 1993; Page 2007; Vanderheiden 2008; Caney 2009).

6. To be clear, we are certain that climate change is occurring and that it is anthropogenically forced (Working Groups I, II, and III 2014, sec. 1.1, 1.2).

7. David Wiens has expressed general pessimism about far-ranging feasibility assessments given their complexity and our epistemic limitations (2015, 20–21).

8. Geoffrey Parker's *Global Crisis* (2013) offers a detailed historical assessment of the relationship between climate change and social, economic, and political upheavals during the Little Ice Age.

9. As Southwood claims, feasibility is often treated as an "agentive modal" (2018, sec. 1.4).

10. Examples include: at the state level, the Least Developed Countries (LDC) Group of the UNFCCC (http://www.ldc-climate.org/); at the sub-state level, the C40 Cities (https://www.c40.org/); in civil society, the Climate Action Network (http://www.climatenetwork.org/).

11. There is reasonable disagreement about substantive normative values and principles that influence rankings of the desirability of institutional transitions.

12. There is reasonable disagreement about the interpretation of evidence bearing on complex empirical questions influencing the assessment of the political feasibility of institutional transitions.

13. Gilabert (2017, 123) engages this concern about disagreement in groups, and offers interesting advice (i.e., focusing on crises, and egalitarian empowerment). But this advice is not related to the probabilistic rational choice model, and we are thus free to accept the advice while rejecting that model.

14. Sayegh offers this example without these specific numbers (2018, 413).

15. LDCs complained that the +2°C threshold of dangerous warming identified by the IPCC may be too high. The Paris Accord captured this complaint by agreeing "to pursue efforts to limit the temperature increase even further [than 2 degrees] to 1.5 degrees Celsius" ("The Paris Agreement" 2020).

16. "Private firms that operate on large spatial scales, have decision-making flexibility, and can weather a poor forecast might in fact be better equipped to make use of forecasts" (Broad and Agrawala 2000).

17. Roser does not engage with the "pathway problem" specifically. I independently apply his arguments in its defense. Any errors in this application are my own.

18. Roser is referring to Bayesian or "subjective" probabilities as degrees of belief in light of information, in contrast to frequentist or classical probabilities as observation of random events (2017, 6–8).

19. This problem seems most pronounced when expected utility maximization remains the principle of choice. Roser does suggest that in relaxing the epistemic demands on probabilities, we might accordingly modify our principle of choice (2017, end of 14).

20. Though, as the IPCC notes, "the assessed uncertainty is larger than the raw model spread" (Working Group III 2014, 175).

21. "It has not, in general, been possible to assign likelihoods to individual forcing scenarios [i.e., RCPs]. ... No probabilities or likelihoods have been attached to the alternative RCP scenarios (as was the case for SRES scenarios)" (Working Group I 2013, 1036, 1038, respectively).

22. Standing contributions, which deserve more discussion than space permits, include Kowarsch and Edenhofer's Deweyan pragmatist "pathway exploration approach" (2016) and David Wiens' "social failure approach" (2015, sec. 5; 2012).

23. The IPCC briefly discusses scenario analysis among other decision-support alternatives (Working Group III 2014, sec. 2.5.7.2, 175).

24. "Different institutions/communities within each Tribe, some of which rarely communicate with each other, were able to share knowledge and insights through storytelling (since scenarios are narratives)" (Whyte et al. 2014, 7).

25. "The third part of the scenario planning workshops involved comparing similarities and differences among each scenario and then brainstorming about current and future Tribal capacities needed to be prepared for these potential scenarios in the future" (Whyte et al. 2014, 5).

REFERENCES

Abeysinghe, Achala, and Saleemul Huq. 2016. "Climate Justice for LDCs Through Global Decisions." In *Climate Justice in a Non-Ideal World*, edited by Clare Heyward and Dominic Roser. Oxford: Oxford University Press.

Agrawala, Shardul, Kenneth Broad, and David H. Guston. 2001. "Integrating Climate Forecasts and Societal Decision Making: Challenges to an Emergent Boundary Organization." *Science, Technology, & Human Values* 26(4): 454–477.

Aldred, Jonathan. 2016. "Emissions Trading Schemes in a 'Non-Ideal' World." In *Climate Justice in a Non-Ideal World*, edited by Clare Heyward and Dominic Roser, 149–168. Oxford: Oxford University Press.

Betz, Gregor. 2009. "Underdetermination, Model-Ensembles and Surprises: On the Epistemology of Scenario-Analysis in Climatology." *Journal for General Philosophy of Science / Zeitschrift Für Allgemeine Wissenschaftstheorie* 40(1): 3–21.

Betz, Gregor. 2010. "What's the Worst Case? The Methodology of Possibilistic Prediction." *Analyse & Kritik* 32(1): 87–106.

Binmore, Ken. 2008. *Rational Decisions*. Princeton: Princeton University Press.

Brandstedt, Eric. 2019. "Non-Ideal Climate Justice." *Critical Review of International Social and Political Philosophy* 22(2): 221–234.

Broad, Kenneth, and Shardul Agrawala. 2000. "The Ethiopia Food Crisis—Uses and Limits of Climate Forecasts." *Science* 289(5485): 1693–1694.

Brown, Seyom. 1968. "Scenarios in Systems Analysis." In *Systems Analysis and Policy Planning: Applications in Defense*, edited by E. S. Quade and W. I. Boucher, 298–310. RAND Corporation. https://www.rand.org/pubs/reports/R0439.html.

Campbell, Kurt M., Jay Gulledge, J. R. McNeill, John Podesta, Peter Ogden, Julianne Smith, Richard Weitz, and Derek Mix. 2007. *The Age of Consequences: The Foreign Policy and National Security Implications of Global Climate Change*. Center for Strategic and International Studies.

Caney, Simon. 2009. "Justice and the Distribution of Greenhouse Gas Emissions." *Journal of Global Ethics* 5(2): 125–146.

Caney, Simon. 2016. "Climate Change and Non-Ideal Theory: Six Ways of Responding to Non-Compliance." In *Climate Justice in a Non-Ideal World*, edited by Clare Heyward and Dominic Roser, 21–42. Oxford: Oxford University Press.

Chermack, Thomas J. 2005. "Studying Scenario Planning: Theory, Research Suggestions, and Hypotheses." *Technological Forecasting and Social Change* 72(1): 59–73.

Clarke, Lee. 2006. *Worst Cases: Terror and Catastrophe in the Popular Imagination*. Chicago: University of Chicago Press.

Dessai, Suraje, Mike Hulme, Robert Lempert, and Roger Pielke Jr. 2009. "Climate Prediction: A Limit to Adaptation?" In *Adapting to Climate Change: Thresholds, Values, Governance*, edited by W. Neil Adger, Irene Lorenzoni, and Karen L. O'Brien, 64–78. Cambridge: Cambridge University Press.

Dryzek, John S. 2016. "Institutions for the Anthropocene: Governance in a Changing Earth System." *British Journal of Political Science* 46(4): 937–956.

Dubois, Didier, and Henri Prade. 2007. "Possibility Theory." *Scholarpedia* 2(10): 2074.

Dubois, Didier, and Henry Prade. 2015. "Possibility Theory and Its Applications: Where Do We Stand?" In *Springer Handbook of Computational Intelligence*, 31–60. Springer.

Farrelly, Colin. 2007. "Justice in Ideal Theory: A Refutation." *Political Studies* 55(4): 844–864.

Friedman, Lisa. 2019. "Trump Serves Notice to Quit Paris Climate Agreement." *The New York Times*. November 4, 2019. https://www.nytimes.com/2019/11/04/climate/trump-paris-agreement-climate.html.

Gardiner, Stephen M. 2006. "A Core Precautionary Principle." *Journal of Political Philosophy* 14(1): 33–60.

Gaus, Gerald. 2011. "Social Contract and Social Choice." *Rutgers LJ* 43: 243.

Gilabert, Pablo. 2017. "Justice and Feasibility: A Dynamic Approach." In *Political Utopias: Contemporary Debates*, edited by K. Vallier and M. Weber, 95–126. Oxford: Oxford University Press.

Gilabert, Pablo, and Holly Lawford-Smith. 2012. "Political Feasibility: A Conceptual Exploration." *Political Studies* 60(4): 809–825.

Hamlin, Alan. 2017. "Feasibility Four Ways." *Social Philosophy and Policy* 34(1): 209–231.

Hamlin, Alan, and Zofia Stemplowska. 2012. "Theory, Ideal Theory and the Theory of Ideals." *Political Studies Review* 10(1): 48–62.

Heyward, Clare, and Dominic Roser. 2016. *Climate Justice in a Non-Ideal World*. Oxford: Oxford University Press.

Houston, Jared. 2020. "Contingency Planning for Severe Climate Change." *Radical Philosophy Review* 23(2): 225–260.

Hsiang, Solomon M., Marshall Burke, and Edward Miguel. 2013. "Quantifying the Influence of Climate on Human Conflict." *Science* 341(6151): 1235367.

Katzav, Joel. 2014. "The Epistemology of Climate Models and Some of Its Implications for Climate Science and the Philosophy of Science." *Studies in History and Philosophy of Science Part B: Studies in History and Philosophy of Modern Physics* 46, Part B (May): 228–238.

Klare, Michael T. 2019. *All Hell Breaking Loose: The Pentagon's Perspective on Climate Change*. New York: Henry Holt and Company.

Kowarsch, Martin, and Ottmar Edenhofer. 2016. "Principles or Pathways? Improving the Contribution of Philosophical Ethics to Climate Policy." In *Climate Justice in a Non-Ideal World*, edited by Clare Heyward and Dominic Roser, 21–42. Oxford: Oxford University Press.

Lawford-Smith, Holly. 2012. "The Feasibility of Collectives' Actions." *Australasian Journal of Philosophy* 90(3): 453–467.

Lawford-Smith, Holly. 2013. "Understanding Political Feasibility." *Journal of Political Philosophy* 21(3): 243–259.

Lempert, Robert. 2013. "Scenarios That Illuminate Vulnerabilities and Robust Responses." *Climatic Change* 117(4): 627–646.

Lempert, Robert J., David G. Groves, Steven W. Popper, and Steve C. Bankes. 2006. "A General, Analytic Method for Generating Robust Strategies and Narrative Scenarios." *Management Science* 52(4): 514–528.

Lempert, Robert J., and Michael E. Schlesinger. 2000. "Robust Strategies for Abating Climate Change." *Climatic Change* 45(3–4): 387–401.

Lloyd, Elisabeth A., and Vanessa J. Schweizer. 2014. "Objectivity and a Comparison of Methodological Scenario Approaches for Climate Change Research." *Synthese: An International Journal for Epistemology, Methodology and Philosophy of Science* 191(10): 2049.

Mastrandrea, M. D., C. B. Field, T. F. Stocker, O. Edenhofer, K. L. Ebi, D. J. Frame, H. Held, et al. 2010. "Guidance Note for Lead Authors of the IPCC Fifth Assessment Report on Consistent Treatment of Uncertainties." *Intergovernmental Panel on Climate Change (IPCC)*. http://www.ipcc.ch.

Moellendorf, Darrel. 2015. "Climate Change Justice." *Philosophy Compass* 10(3): 173–186.

Moore, Sara S., Nathaniel E. Seavy, and Matt Gerhart. 2013. "Scenario Planning for Climate Change Adaptation: A Guidance for Resource Managers." Point Blue Conservation Science and California Coastal Conservancy.

Ndikumana, Léonce, and James K. Boyce. 2011. *Africa's Odious Debts: How Foreign Loans and Capital Flight Bled a Continent*. London: Zed Books Ltd.

Oels, Angela. 2013. "Rendering Climate Change Governable by Risk: From Probability to Contingency." *Geoforum* 45(March): 17.

Page, Edward A. 2007. *Climate Change, Justice and Future Generations*. Cheltenham: Edward Elgar Publishing.

Parenti, Christian. 2011. *Tropic of Chaos: Climate Change and the New Geography of Violence*. New York: Nation Books.

Parker, Geoffrey. 2013. *Global Crisis: War, Climate Change and Catastrophe in the Seventeenth Century*. New Haven: Yale University Press.

Peterson, Martin. 2009. *An Introduction to Decision Theory*. Cambridge Introductions to Philosophy. Cambridge: Cambridge University Press.

Rawls, John. 1999. *The Law of Peoples: With, The Idea of Public Reason Revisited*. Cambridge, MA: Harvard University Press.

Rickards, Lauren, John Wiseman, Taegen Edwards, and Che Biggs. 2014. "The Problem of Fit: Scenario Planning and Climate Change Adaptation in the Public Sector." *Environment and Planning C: Government and Policy* 32(4): 641–662. https://doi.org/10.1068/c12106.

Robeyns, Ingrid. 2008. "Ideal Theory in Theory and Practice." *Social Theory and Practice* 34(3): 341–362.

Roser, Dominic. 2017. "The Irrelevance of the Risk-Uncertainty Distinction." *Science and Engineering Ethics* June: 1–21.

Sayegh, Alexandre Gajevic. 2018. "Justice in a Non-Ideal World: The Case of Climate Change." *Critical Review of International Social and Political Philosophy* 21(4): 407–432.

Schoemaker, Paul JH. 1995. "Scenario Planning: A Tool for Strategic Thinking." *Sloan Management Review* 36(2): 25–50.

Schwartz, P., and D. Randall. 2004. "An Abrupt Climate Change Scenario and Its Implications for United States National Security." *Oil, Gas & Energy Law Journal (OGEL)* 2(1).

Shue, Henry. 1993. "Subsistence Emissions and Luxury Emissions." *Law & Policy* 15(1): 39–60.

Simmons, J.A. 2010. "Ideal and Nonideal Theory." *Philosophy & Public Affairs* 38(1): 5–36.

Southwood, Nicholas. 2018. "The Feasibility Issue." *Philosophy Compass* 13(8): e12509.

Spratt, David, and Ian Dunlop. 2019. "Existential Climate-Related Security Risk: A Scenario Approach." Breakthrough – National Centre for Climate Restoration.

Stemplowska, Zofia. 2016. "Feasibility: Individual and Collective." *Social Philosophy and Policy* 33(1–2): 273–291.

Tsebelis, George. 1989. "The Abuse of Probability in Political Analysis: The Robinson Crusoe Fallacy." *The American Political Science Review* 83(1): 77.

United Nations|Climate Change. 2020. "The Paris Agreement." https://unfccc.int/process-and-meetings/the-paris-agreement/the-paris-agreement.

Valentini, Laura. 2009. "On the Apparent Paradox of Ideal Theory." *Journal of Political Philosophy* 17(3): 332–355. https://doi.org/10.1111/j.1467-9760.2008.00317.x.

Valentini, Laura. 2012. "Ideal Vs. Non-Ideal Theory: A Conceptual Map." *Philosophy Compass* 7(9): 654–664.

Vanderheiden, Steve. 2008. *Atmospheric Justice: A Political Theory of Climate Change*. Oxford: Oxford University Press.

Whyte, K. P., M. Dockry, W. Baule, and D. Fellman. 2014. "Supporting Tribal Climate Change Adaptation Planning Through Community Participatory Strategic Foresight Scenario Development." Project Reports. Great Lakes Integrated Sciences and Assessments (GLISA) Center.

Wiens, David. 2012. "Prescribing Institutions Without Ideal Theory." *Journal of Political Philosophy* 20(1): 45–70.

Wiens, David. 2015. "Political Ideals and the Feasibility Frontier." *Economics & Philosophy* 31(3): 447–477.

Working Group I to the Fifth Assessment Report of the Intergovernmental Panel on Climate Change. 2013. *Climate Change 2013: The Physical Science Basis*. Cambridge: Cambridge University Press.

Working Group III to the Fifth Assessment Report of the Intergovernmental Panel on Climate Change. 2014. *Climate Change 2014: Mitigation of Climate Change*. Cambridge: Cambridge University Press.

Working Groups, I, II, and III to the Fifth Assessment Report of the Intergovernmental Panel on Climate Change. 2014. *Climate Change 2014: Synthesis Report*. Cambridge: Cambridge University Press.

Chapter 7

Making the Great Climate Transition
Between Justice and Feasibility
Fabian Schuppert

If the currently living generation wants to have any chance of avoiding the most dangerous consequences of anthropogenic climate change, we need to change the way people live.[1] This is particularly true for the vast majority of people living in the so-called Global North. Per capita greenhouse gas emissions and overall resource consumption in these countries is way above sustainable levels (Hoekstra and Wiedmann 2014; Rosales 2008). This means that the required changes in how people live, how natural resources are extracted and used, how goods are produced and consumed, and how people live and travel will have to be drastic. Put differently, any hope of mitigating the worst aspects of climate change requires radical change, especially among the affluent inhabitants of the Global North. This insight is often repackaged as a conversation about "climate transition," "sustainable transition," "the green new deal," "green economy," "sustainable development," or a "green future."[2]

Climate transition comes with a whole range of costs and possible benefits, whether it is the creation of new jobs in transition-relevant industries or the loss of jobs in industries that are deemed transition-incompatible. This raises obvious issues of distributive justice. It is thus not surprising that debates over climate transition latched onto the discourse on "just transition," a framework that was originally developed by the trade union movement to protect the interests of workers who might stand to lose their jobs through increased environmental protection measures. The basic assumption of the just transition literature is that the major socio-economic changes that the climate transition will necessitate should affect people in a fair and equitable manner (Green and Gambhir 2020; Piggot et al. 2019; McCauley and Heffron 2018). This seems spot on: since socio-economic transformation bears the risk of exacerbating existing injustices and creating new ones, justice should be (at least) one of the relevant yardsticks for identifying an appropriate climate transition pathway.

However, any conversation about the kind of climate transition needed quickly turns into a debate over what kind of changes are feasible, since it is commonly taken for granted that it would be a waste of time to discuss changes that aren't feasible. As Pickering et al. (2012, 427) observe "if we are concerned about fairness, we will have good reason to be concerned about feasibility . . . [but] a key limitation of many proposals for fairly distributing the benefits and burdens of mitigation—both in climate ethics and climate economics—is a tendency to pay limited attention to their feasibility."

In other words, justice and fairness are key normative considerations for designing transition pathways, but unless one properly factors in feasibility, this work might be in vain and of limited practical use. As Eric Posner and David Weisbach (2010) put it, an effective global climate treaty is only possible if one focuses on what makes a proposal for all actors feasible and so we must leave considerations of justice for another day. On this view, feasibility is a major constraint on the range of possible options and the assessment of feasibility seems to come prior to the assessment of justice, if justice plays any role at all. As Dominic Roser (2015, 72) points out, this view might be expressed in the injunction, "Choose the least unjust option from the feasible set."

However, as recent contributions to the debate over feasibility have shown (Roser 2015, 2016; Southwood 2016; Lawford-Smith 2013), this might be a bit too quick. First of all, one needs to address the obvious question of what the proper relationship between justice and feasibility on the level of theory-building is and whether it makes sense to engage in the exercise of what is commonly called "ideal theorizing." There has been a lively debate on this in political and moral philosophy (Gheaus 2013; Miller 2013; Gilabert and Lawford-Smith 2012), showing that feasibility should not be a hard constraint on ideal theorizing (e.g., Gilabert and Lawford-Smith 2012). Second, one needs to investigate what it actually means for something to be feasible and who is in the position to decide whether a course of action is feasible or not. This is the more applied side of the debate, largely happening in political theory (Heyward and Roser 2016; Räikkä 1998; Lawford-Smith 2013) and normative policy discourses (Baatz 2018; Skodvin et al. 2010), where feasibility arguments play a major role in deciding which policy option to pursue and for which reasons. It is precisely this second set of questions which lies at the heart of this chapter.

The chapter is comprised of four sections. The first section will briefly look at existing research on feasibility, which identifies at least two different kinds of feasibility. As recent contributions to the debate have shown, most forms of feasibility are non-binary, socially mediated assessments of what is likely to succeed in a given context. The next section, then, will engage in a form of ideology critique and highlight the way that feasibility perceptions are interlinked with existing power structures and forms of privilege. By uncovering this connection one can expose a large range of feasibility arguments that are actually morally unjustifiable. Following, I will argue for prioritizing minimizing injustices in climate transition, precisely because of existing inequalities in power, voice, and influence. I conclude by arguing that the proper place of feasibility arguments in the climate transition debate is in the political sphere where they shape discussion of which sacrifices can reasonably be demanded of other people.

I. FEASIBILITY: ITS KINDS AND FUNCTIONS

In the past few years, the idea of feasibility has received a lot of attention in normative political theory and philosophy, both in more abstract discussions of how theorists should go about building normative theories and in more applied discussions concerning politics and policymaking. The focus of this chapter is on the latter, as one of its aims is to assess the proper normative status of many feasibility arguments in the climate transition debate. In order to fulfill this aim, it makes sense to take one's bearings from the existing literature, which distinguishes very roughly between at least two types of feasibility.

First, feasibility can function as a binary statement about absolute possibility (what Gilabert and Lawford-Smith (2012) call binary feasibility). Infeasible in this context simply means impossible. The objective probability for x working as planned is zero. Any solution with a probability greater than zero is in this sense feasible, even if there might be further thresholds with probabilities greater than zero, such as, for instance, "likely" and "unlikely" for probabilities that are greater/lower than a predetermined threshold level y.

This type of feasibility is often used in the context of technological feasibility. Those working in the natural and technological sciences might assume that technological feasibility is a binary and objective matter: either something is technologically possible, and hence feasible, or it is not, and one can ascertain what is the case in an objective, uncontroversial way. Reality, however, often looks different. As the case of large-scale geoengineering shows, even what counts as technologically feasible can be widely contested. In addition, many models that are used to determine whether something counts as feasible are in need of greater scrutiny, as in the case of integrated assessment models, which have become fairly authoritative when it comes to charting "feasible" transition pathways (Low and Schäfer 2020). This brings us to the second group of feasibilities.[3]

Second, feasibility assessments can be statements about where, on the wide spectrum ranging from "merely possible" to "extremely likely," a particular option ranks (Lawford-Smith 2013, 256). Virtually all statements of what is economically, socially, or politically feasible fall into this category. Economic feasibility refers to whether the costs for a proposed solution are justifiable, proportionate, and manageable. Social feasibility refers to whether a proposed change stands any chance of getting social approval, that is, whether people will be willing to comply with certain norms, adopt certain practices, or accept certain changes. Then there is political feasibility, which has received a fair amount of attention in the last few years (Gilabert and Lawford-Smith 2012; Jewell and Cherp 2020). Political feasibility refers to whether something is likely to be passed as a law, or will be successful at the polls.

Feasibility statements of this second kind are often used to express the idea that something is overly costly, socially unpopular, or (not) very likely to be politically successful. Both in academic as well as policy debates, this second kind of feasibility is often presented as a kind of "reality check" for overly utopian and ambitious proposals. The basic idea behind this second kind of feasibility is that one needs to be "realistic" about what might be possible in the future, considering what we know about the world as it is and the kinds of motivations and dispositions people and institutions have.

However, what counts as economically, socially, or politically feasible is not fixed but shaped by social conditions. That is one of the key insights of the existing literature on economic, social, and political feasibility (Lawford-Smith 2013). Emily McTernan (2019), for instance, points out that even if we use all the insights from social sciences available to us, we would not be able to formulate accurate, reliable, and non-arbitrary feasibility standards. The reason for this is that we cannot claim that what we know about people's preferences and dispositions today provides us sufficient certainty about what will happen in the future if a particular policy option is being tabled. Put differently, when it comes to assessing an option's feasibility, it is not usually as easy as looking at some empirical findings, reading off the facts, and having a firm idea of what needs to be the case for the option to be feasible.

In the existing normative literature, feasibility has been usefully problematized by showing the notion's internal complexities. First of all, most of the time when feasibility is introduced as a consideration it is meant as a socially mediated, somewhat malleable category. Second, because feasibility normally refers to a malleable, socially mediated category, one needs to ask important questions: feasible for whom, with regard to what, where, and when? Asking these kinds of questions helps in identifying how fixed certain feasibility constraints are (Gilabert and Lawford-Smith 2012, 812). It also allows one to identify a range of mistakes and distortions of which feasibility claims may be guilty. As Roser (2015, 73–80) observes, many feasibility claims lack empirical justification, fail to explain the considerations that make an option infeasible and fail to identify the normative weight of those considerations, fail to relativize their assessment with regard to time and place, and use current assumed motivation as a proxy for future behavior and actions. However, only when one knows for which agent (where and at what point in time) a particular X is infeasible and for which reasons, can one arrive at a judgment on whether the presented (in)feasibility claim is indeed justified.

In other words, feasibility claims are complex claims based on several criteria (who, where, when, with regard to what?), and need to be based on reasons that are clear and accessible. So there are two sides to any critical engagement with feasibility, understood as a socially malleable, non-binary category: first, one needs to look at who and what the feasibility claim applies to, in which circumstance, and when; second, one needs to look at the justificatory logic of each feasibility claim. The latter aspect refers to the kind of reasons on which feasibility claims can be based. There are many convincing theories of how to identify said reasons (e.g., reasons based on common avowable interests, reasons based on the basic values of our democratic culture, etc.) and it does not seem necessary to commit oneself too strongly to one particular theory. However, since this chapter is concerned with the normative status of feasibility claims in the climate transition discourse, we can stipulate that all feasibility claims ought to satisfy at least some basic normative criteria in order to be deemed defensible: feasibility claims need to be based on accessible reasons that satisfy at least a minimal understanding of public reason,[4] and so they should not be based on claims that are inherently unjust or overtly discriminatory. Asking this much should hopefully not be particularly controversial.

In short, as with any other claim, feasibility claims need to be carefully assessed and one needs to determine whether a particular feasibility claim is indeed justifiable and whether one should attach any kind of normative weight to it. As the following sections will show, many feasibility claims are not justifiable. Some have argued that the reason for this is that many feasibility claims are grounded in spurious arguments based on generalizing claims about people's moral motivation, or the general political mood in a country, both of which require fairly complex empirical backing in order to be fully convincing (McTernan 2019; Roser 2015). I will argue that a further reason that many feasibility claims are unjustifiable is that they are intricately linked with existing power imbalances, structural injustices, and unwarranted privilege.

II. FEASIBILITY, POWER, AND PRIVILEGE

Virtually all feasibility claims are status quo based, that is, they take the status quo as baseline from which to judge whether a particular proposal is feasible or not. Take for instance the debate over whether all existing housing should be retrofitted to be made more climate efficient. The feasibility assessment of this proposal commonly depends to a large degree on what the status of exiting housing is, what the current prices of building materials and labor are, and a range of expected side-effects, which are, in most cases, calculated based on current indicators and standard modeling (Jones et al. 2013; Mata et al. 2019). However, at closer analysis it becomes clear that feasibility claims can be, in two very different ways, status quo based. First, feasibility claims may take the status quo as a presumably neutral given. In this case, the fact that we are in state of affairs Y is simply taken to shape what counts as feasible and what doesn't. There is no critical questioning of the status quo, of the rights and duties that shape the status quo, and the historical context around the status quo. It doesn't matter how we got to state of affairs Y or how unjust it is, it only matters what people's attitudes are, how much a proposal like retrofitting costs, and what the expected effects of such policies might be. Alternatively, feasibility claims can be based on an in-depth critical analysis of the status quo, uncovering not just the socially mediated and malleable nature of different kinds of parameters that influence a proposal's feasibility (such as people's attitudes, the proposal's political support, and so on), but also unmasking how existing power structures and social inequalities might shape the very basic data points that go into any feasibility assessment.

In the context of the climate transition, many obvious proposals are routinely declared to be off the table. Charge commercial aviation's elite super emitters for their CO_2 emissions (Gössling and Humpe 2020)? A paradigmatic case of the politics of envy and blind activism. Restrict or heavily tax the use of high-emitting luxury goods (Benoit 2020)? Undue interference in people's privacy, which will never receive political support. But why is this? And why are other proposals deemed at least worthy of discussion?

When it comes to assessing a proposal's feasibility not all proposals are created equal. Therefore, the discourse around the climate transition needs to critically challenge the very ideas and concepts that provide the conceptual frame for feasibility claims in the status quo and from there identify which (if any) feasibility claims can be morally justified. This moral assessment needs to include an in-depth look at how the feasibility claim in question relates to the status quo and existing injustices and inequalities within the status quo. This is because feasibility claims stemming from partially or fully unjust circumstances should be subjected to particularly critical normative scrutiny.

This is necessary in order to make sure that the feasibility assessments are not complicit in upholding an unjust system.

In the policy-oriented literature on feasibility (Spencer et al. 2018; Nielsen et al. 2020), and in debates over so-called feasibility constraints in practice,[5] a policy is only framed as feasible if people are not forced to make too large sacrifices with regard to their current standard of living, well-being, or economic expectations. Yet, the fact that there are currently massive distributive injustices in standards of living, well-being, or economic expectations is not seen to affect the validity of that feasibility assessment. In other words, feasibility as a reality check on proposals that demand too large sacrifices from people, is introduced into the debate as an obvious constraint derived from looking at a seemingly neutral baseline, namely, the world as it is, including people's current property holdings and their attitudes toward having less.

However, the world as it is, is not just. In fact, it is ripe with injustices, many of which are directly linked to the phenomenon of climate change. There is ample literature on the per capita resource overconsumption by people in the Global North, both past and present, the key role that a handful of industries and a limited number of corporate agents play in the emission of greenhouse gases, and the neo-colonial nature of many proposed climate mitigation and adaptation strategies (e.g., Lyons and Westoby 2014; Mahony and Endfield 2018). On top of that, many of today's existing, morally problematic inequalities can be traced back to obvious injustices such as colonialism, imperialism, racism, and capitalist resource exploitation. In light of all of these observations, it is clearly impossible to take the status quo as a neutral baseline, even if one thinks that assigning responsibility for past wrongs and the wrongs connected to climate change might be impossible or undesirable.[6] Being unable to assign moral responsibility and liability does not mean that existing practices are morally unproblematic or that one can straightforwardly form legitimate expectations based on these practices.

As the literature on both male and white privilege shows (Sullivan 2006, 2019; Frye 1983; McIntosh 1989; Manne 2020), being unintentionally in, and non-blameworthy for, a position of privilege does not render said privilege morally unproblematic. In fact, because these privileges in part stem from both historical and structural injustices, agents in a position of privilege might well have a strong moral duty to act differently. This is precisely what the argument of this chapter is. Building on the work of Iris Marion Young (2003, 2009), as well as feminist work and critical race theory, scholars working in the areas of climate change and climate transition should look at the phenomenon of structural injustice and how it shapes *what* counts as feasible and *who* determines this. As this analysis will show, many nontechnical feasibility claims are based on habits of privilege and the structural power imbalances that sustain existing hierarchies and keep the privileged where they are: on top.

Taking one's bearings from Jeremy Dunham and Holly Lawford-Smith (2017, 5–6), who in turn build on the trailblazing work by Peggy McIntosh (1988, 2012) and Alison Bailey (1998), for the purposes of the analysis here we can define privilege quite broadly as "a species of advantage" (Dunham and Lawford-Smith 2017, 6) that is undeserved and "conferred systematically to members of [particular, broadly identifiable] groups" (Dunham and Lawford-Smith 2017, 6). Furthermore, one's privilege is arbitrary because one has it simply as a result of one's group membership. Privilege is also inherently connected to the disadvantages that others suffer and thus a form of unequal power or advantage that is "almost never justifiable" (Bailey 1998, 108). This connection between privilege and power was what led Peggy McIntosh to formulate her iconic account of privilege as "unearned power conferred systematically" (1988, 3). How, then, does all this connect to the discussion of feasibility claims and climate transition?

As pointed out earlier, many feasibility claims refer in problematic ways back to the status quo, without any critical analysis of the supposed facts, or worse with a thinly veiled form of ignorance and bias. Take the example of global climate mitigation through the large-scale use of bioenergy with carbon capture and storage (BECCS). While BECCS is generously funded by several European countries, BECCS deployment in Europe is very low. This almost certainly has something to do with policymakers failing to create the right incentives and providing a reliable framework for BECCS deployment (Geden et al. 2018; Fridahl et al. 2020). However, there is another important element.

Europe is seen as an infeasible site for BECCS deployment because of existing property holdings, the current use of land by highly productive agri-businesses and farmers, as well as the associated social costs of shifting demands on farmers that will force them to partially change the way they work. In addition, deployment elsewhere is assumed to be cheaper, especially in terms of social costs, as it is assumed that in other parts of the world, in particular sub-Saharan Africa, large swaths of land exist that are just waiting to be used for large-scale BECCS deployment (Hayashi et al. 2020). Surprisingly, the current land use of local farmers and the social costs in other parts of the world are hardly ever mentioned in this context. Similarly, the need for a strong and reliable policy framework for large-scale BECCS implementation, not least because of environmental concerns, hardly gets mentioned when the focus of the studies/discourses is on sub-Saharan Africa.

What one sees at play here is, first, belief that privileged people in the Global North can object to certain changes if those changes impact their lifestyles and property rights, and second, a deeply colonial (and possibly racist) attitude that people in "poor" or "underdeveloped" countries ought to be happy to take up opportunities like BECCS and to help mitigate global climate change. Similarly, the idea that people from the Global North should donate money to plant lots of trees in sub-Saharan Africa, irrespective of the questionable scientific credentials of such proposals, highlights problematic assumptions about the Global South as such. On top of that, the tree-planting proposal is normally presented as an "effective and easy" way to solve climate change, which requires no major changes to the way people live, produce, and consume in the Global North.

Thus, many feasibility arguments implicitly suggest that the status quo with all its features and consumption patterns is something worth preserving, as if it were inherently negative to change the status quo, even though the world we live in is blatantly unjust, both in terms of current practices as well as its genealogy. Many of the unjust features of the world are structural or systemic in nature, and often based on a combination of past injustices leading to normalized inequalities in the present. One particular effect of this working together of past and present injustices is that members of particular groups enjoy specific privileges, such as male privilege, white privilege, ableist privilege, and so on. These different kinds of privilege often overlap and intersect.

One often overlooked form of privilege is environmental privilege, which can be aptly described as the opposite of the harmful environmental inequalities members of other groups experience (Murphy 2016). It also refers to privileged access to certain environmental goods, such as taking up of greater ecological space as a result of engaging in putatively essential activities, and to freedom from environmental burdens (or significantly reduced vulnerability, as is often the case in the context of anthropogenic climate change). Environmental privilege is an extremely useful concept for identifying the kind of advantages that certain (privileged) groups enjoy when it comes to scarce environmental resources and (relative) sheltering from the negative effects of environmental degradation. Just as with other forms of privilege, many people who are environmentally privileged fail to realize that they are. Especially when it comes to climate change, which is chiefly framed as an issue of risk and vulnerability, the distinct absence of that vulnerability, which is a form of environmental privilege, just doesn't register with the privileged. For them, it is simply normal not to have to worry about water shortages, bad harvests, or loss of biodiversity. Unsurprisingly, environmental privilege in part traces the markers of other forms of privilege such as race, socio-economic status, or geographical location (Murphy 2016). In countries like the United States of America, for instance, adaptation planning in coastal areas has partially reinforced racial injustices and disadvantage (Hardy et al. 2017).

Apart from the kinds of environmental advantages mentioned earlier, which obviously play a major role in the context of anthropogenic climate change, privilege brings a whole host of other advantages. More often than not privilege goes hand in hand with three problematic phenomena: unjustifiably increased social recognition, inflated epistemic authority, and a socially accepted sense of entitlement. Closer examination of these three phenomena sheds light on the central questions raised earlier: How does privilege shape judgments about *what* counts as a legitimate feasibility claim and *who* decides that?

First, privilege commonly manifests itself in part through unjustified increased social recognition. While much of the recognition literature focuses on misrecognition and the lack of recognition of those who are socially marginalized and oppressed (Honneth 1995; Fraser and Honneth 2003; Taylor 1994), this "deficit model" only tells half of the story. As Cillian McBride (2013) observes, it is just as morally problematic when individuals and groups (demand and) get an excess share of social recognition without good reason or proper justification, but rather based on structural and habitual recognition ascriptions. This unjustified recognition surplus makes it more likely for the esteemed to occupy positions of power or to demand greater attention and respect. This directly ties in with the second problematic phenomenon.

Second, privilege manifests itself in inflated epistemic authority. As with the more general recognition literature, much of the literature around epistemic injustice was initially concerned with those who get silenced, the testimonially disadvantaged or the hermeneutically deprived. This was for good reason, as these issues are deeply morally problematic (Fricker 2007; Kidd et al. 2017; Dotson 2011; Bhargava 2013). However, the flip side is the fact that members of privileged groups often enjoy undue epistemic standing and unjustified epistemic authority (Medina 2011). This inflated epistemic authority unduly narrows the playing field for other views, facilitates hegemonic silencing, and translates into direct influence over a whole range of things, such as the setting of norms, or the assessment of proposals. Having unjustified increased social recognition and inflated epistemic authority, then, makes it much easier for privileged agents to shape particular discourses, such as (for instance) the debate over what counts as a justifiable feasibility constraint. In other words, if one wants to investigate *who* shapes feasibility discourses, one should look at the kind of influence exerted by privileged groups, and the fights over power and recognition that structure many mitigation and adaptation debates (Nightingale 2017).

Third, because privilege is a systemic and structural phenomenon, and thus a feature deeply ingrained into the social and political workings of society, privilege is most often not identified as such. Habits of privilege are perceived to be "normal" or "legitimate" meaning that the unjustified advantages that privileged members of society receive are translated into socially accepted claims. Upon closer look, though, these claims are nothing more than an unjustifiable sense of entitlement that is propped up by unjust structures and pervasive ignorance (Sullivan and Tuana 2007; Manne 2020). Once this link between privilege and unjustified entitlement, which masquerades as legitimate claims, becomes clear, one can critically challenge the often implicitly made assumption that certain feasibility claims must be legitimate because they are based on the status quo.

The fact that the current system is deeply unjust, and that climate change itself is a major injustice (and an expression of colonial power), must be seen to play a significant role in determining what counts as a morally justifiable feasibility concern. When it comes to assessing the validity of different feasibility arguments, and especially those that are built on assumptions that a particular aspect of the status quo is inherently worth preserving, one needs to carefully analyze said argument, check how it connects to existing injustices, and assess whether it is an expression of undue privilege. The fact that the person or institution putting forth the argument might well be ignorant about their own or somebody else's sense of entitlement does not mean that claims based on habits of privilege and entitlement should be socially accepted. In fact, the opposite is the case: habits of privilege need to be unmasked, and the influence and epistemic authority of privileged groups in discussions over feasibility need to be challenged and overcome.

III. FIGHTING INJUSTICE FIRST, WHILE NOT LOSING SIGHT OF THE OVERALL GOAL

The climate transition will come with a range of difficult choices, many of which might exacerbate existing inequalities and injustices. Therefore, any climate transition should strive to be just or at least justice-promoting.[7] However, in order for climate transition efforts to be anywhere near this ideal, the proper place and the adequate normative weight of different feasibility claims needs to be specified.

As argued earlier, the fact that the world is already unjust and that climate change is an injustice itself, often with clearly colonial undertones, is relevant to the project of achieving a just climate transition. If the idea that the current baseline is neutral is jettisoned, then it becomes clear that the first step to accomplish any kind of "just" transition is to identify existing injustices and to do two things: first, look at how these existing injustices are tied (if at all) to the problem at hand, i.e., the massive harms connected to anthropogenic climate change; second, look at how these existing injustices shape what is perceived to be feasible and so act as blocks to certain avenues of necessary change.

If one does so, one can observe that the pronounced geographies of risk and vulnerability connected to climate change, how vulnerability and exposure to risk are distributed across populations and communities, are—at least to a significant degree—connected to issues of past and present systemic and structural injustice. These past and present injustices are often part of the problematic norms and behaviors that led to global anthropogenic climate change in the first place. In addition, many of these injustices unduly shape the charting of possible transition pathways.

So instead of rushing to give normative weight to the privilege-based claims from feasibility that dominate much of the climate transition discourse, one should focus on analyzing the normative credentials of these claims. Since privilege-based feasibility claims will exacerbate rather than reduce existing inequalities and injustices, one should ignore these claims and instead try to identify mitigation and adaptation strategies that also reduce existing injustices, while not creating new ones. If transition pathways follow this route, they will actually fight injustices in two ways: first, by addressing existing injustices; second, by addressing the very injustice that is climate change.

Needless to say, in many cases this might be easier said than done. Is, for instance, the blocking of lower speed limits on motorways in Germany (i.e., the infamous *Autobahn*) a claim that is based on privilege, since it is mainly the well-off in Germany that can afford to buy cars that can take advantage of the "no speed limit" zones, or is it a claim that is based on the narrow self-interest of powerful lobby groups connected to the car industry? Or is it simply that many Germans value this freedom so much that they are not willing to sacrifice this "perk" of using German motorways? It might well be that the answer is a mixture of all three suggested explanations. So how will one know what the correct answer is and which feasibility claims should be discounted and which are justified?

This is where the proper place of feasibility claims comes in. As mentioned earlier, the vast majority of feasibility claims cannot be treated like a hard constraint on available options, or even a way to limit the available option set. Feasibility arguments operate mostly at a lower level. Feasibility arguments are arguments about weighing different values and scenarios, as well as making predictions about people's behavior, motivations, and so on. However, because we know—as I established earlier—that many feasibility claims operate with questionable assumptions about the supposed neutrality of the status quo and might even be directly based on morally unjustifiable privilege, no matter which pathway one ultimately chooses, one needs to do two things: first, address existing injustice that has given different groups and individuals undeserved advantage in shaping what pathways are judged as insufficiently feasible; second, critically challenge the supposed limits of the feasibility debate by looking carefully at why certain arguments are put forth and who would benefit from a proposed policy, even if it was only by avoiding a more costly option. Only once these two steps are taken should one present feasibility arguments regarding how values should be weighed and balanced ought to operate.

Despite all the criticisms of certain feasibility claims, which were discussed earlier, there are of course feasibility claims that do merit proper discussion. The proper place for doing so is public socio-political discourse. That socio-political discourse will have to follow certain norms, such as norms of epistemic justice (Fricker 2007) and being non-discriminatory. What matters is that the mere fact that people might be reluctant to act on insight X does not in and of itself count as a weighty reason not to pursue policy Y that follows from X. One first needs to know where that reluctance comes from and who it applies to and why. There might be a significant normative difference between the claim of the owner of a cement factory in Germany who is unwilling to come to terms with their business's environmental impact and the owner of a market stall in Lima who is worried that they won't be able to get a sustainable supply of produce. Whether there is such a difference and how it affects the normative standing of each claim depends on the reasons that can be given in their support, as well as the assumptions that they are based on.

IV. CONCLUSION

This chapter addressed the question of how concerns about feasibility and justice should feature in the debate over the climate transition. While feasibility is often presented as a major constraint that allows people to identify the range of options to choose from, the analysis in this chapter followed recent advances in the debate on social and political feasibility that suggest a very different role for feasibility. Based on the commonly made distinction between different kinds of feasibility, I argued that most feasibility constraints are not only socially mediated and thus malleable factors that should not enjoy lexical priority in determining the range of available options, but also that many arguments from feasibility operate with a deeply problematic assumption regarding the status quo, namely that it provides us with a neural baseline to start from. For many feasibility claims this assumption proves problematic since taking the status quo as a neutral starting point can and often does lead to a problematic masking and perpetuation of existing injustices.

In this chapter, I focused on those feasibility claims that are based on privilege. Privilege is often perceived as being normal or natural, which is precisely one of the reasons why it so easily happens that feasibility claims asserted from a place of privilege exert unjust influence over the climate transition debate. Proposals that seem to challenge current resource and property holdings are simply deemed too unrealistic and infeasible to be further discussed. However, this is deeply problematic. Since the status quo is structurally unjust, and climate change itself is an injustice, too, it is of the utmost importance that one, first, addresses existing injustices that exclude certain voices from the very debate, and second, critically challenges the supposed limits of the feasibility debate by looking carefully at why certain arguments are put forth and who would benefit from a proposed policy. These steps are critical to having any chance of achieving a just climate transition.

Hence, I argued that fighting injustices in the status quo and avoiding new injustices in the process of transformation should take priority over a focus on feasibility. While feasibility is clearly normatively important, since the (futile) striving for an infeasible ideal endpoint should not get in the way of making a positive difference on the ground, the proper place of feasibility is where all social and political disagreements over the proper weighing of values should go: within the socio-political discourse that follows certain norms, such as norms of epistemic justice (Fricker 2007) and critical civility (Whitten 2021). To give most feasibility claims greater normative standing would enshrine unjust privilege into the very process of trying to make a just climate transition. This is precisely what one needs to avoid, if one wants to achieve a truly just transition.

NOTES

1. Acknowledgements: This chapter greatly benefitted from the critical feedback provided at the research colloquia at Free University Berlin, the University of Münster, and the University of Hamburg. I would particularly like to thank my hosts for their invaluable feedback and for providing me with the opportunity to present my (at that point) barely above the embarrassment-threshold operating work-in progress: Franziska Dübgen, Stefan Gosepath, Tamara Jugov, and Peter Niesen. I also owe special gratitude to Katja Wiesner for research assistance, and to Sarah Kenehan and Corey Katz for their patience, encouragement, and comments. Furthermore, I would like to thank Svenja Ahlhaus, Christian Baatz, Valentin Beck, Daniel Häuser, Matthias Hoesch, Gesche Jeromin, Felix Koch, Ervin Kondakciu, Dominic Lenzi, Luise Müller, Esther Neuhann, Markus Patberg, Carmen Puchinger, Maximilian Sarre, Sophia Schulze-Schleithoff, Tobias Schweitzer, for their insightful questions, which helped me to improve the chapter.

2. For the remainder of this chapter I will use the term "climate transition" to refer to the large scale socio-economic changes that need to happen in order to safeguard

well-being and livelihoods while also transitioning to a more sustainable, climate change mitigating socio-economic system.

3. For the remainder of the chapter, however, I will reserve the term "technical feasibility" for feasibility claims of the first kind, i.e., those referring to objective possibility.

4. The idea of public reason is obviously a fairly complex one. However, for the purposes of this chapter it is possible to operate with a variety of conceptions of public reason. Considering that later parts of the argument focus on privilege and misrecognition, a suitable conception probably ought to include a substantial commitment to epistemic justice and the rejection of structures of discrimination and domination. I am grateful to Peter Niesen and Stefan Gosepath for encouraging me to include this point.

5. In the actual debate over climate transition pathways, feasibility arguments are regularly framed as constraints. Philosophically speaking most feasibility arguments are not proper "hard" constraints, as Gilabert and Lawford-Smith (2012) have already pointed out. Thank you to Felix Koch for pressing me on this point.

6. While I actually do not share this view as a whole, it is important to stress that the argument presented here still holds even if one subscribes to this view.

7. Obviously, choosing an appropriate account of justice is no easy task. However, for the purposes of this chapter I take it for granted that any climate transition can only be considered just if it is broadly along social egalitarian lines, since that includes the addressing of certain long-standing structural and systemic inequalities, especially along lines of race, gender, and sexual orientation. However, the arguments presented here are broadly compatible with a range of other conceptions of justice, as most existing theories do take issue with certain forms of privilege.

REFERENCES

Baatz, Christian. 2018. "Climate Adaptation Finance and Justice: A Criteria-Based Assessment of Policy Instruments." *Analyse & Kritik* 40(1): 73–106.

Bailey, Alison. 1998. "Privilege: Expanding on Marilyn Frye's 'Oppression.'" *Journal of Social Philosophy* 29(3): 104–119.

Benoit, Philippe. 2020. "A Luxury Carbon Tax to Address Climate Change and Inequality: Not All Carbon Is Created Equal." *Ethics and International Affairs*. https://www.ethicsandinternationalaffairs. org/2020/a-luxurycarbon-tax-to-addr ess-climate-change-and-inequality-not-all-carbon-is-created-equal.

Bhargava, Rajeev. 2013. "Overcoming the Epistemic Injustice of Colonialism." *Global Policy* 4(4): 413–417.

Dotson, Kristie. 2011. "Tracking Epistemic Violence, Tracking Practices of Silencing." *Hypatia* 26(2): 236–257.

Dunham, Jeremy, and Holly Lawford-Smith. 2017. "Offsetting White Privilege." *Journal of Ethics & Social Philosophy* 11(2): 1–22.

Fraser, Nancy, and Axel Honneth. 2003. *Redistribution or Recognition?: A Political-Philosophical Exchange*. London: Verso.

Fricker, Miranda. 2007. *Epistemic Injustice: Power and the Ethics of Knowing.* Oxford: Oxford University Press.

Fridahl, Mathias, Rob Bellamy, Anders Hansson, and Simon Haikola. 2020. "Mapping Multi-Level Policy Incentives for Bioenergy with Carbon Capture and Storage in Sweden." *Frontiers in Climate* 2: 25.

Frye, Marilyn. 1983. *The Politics of Reality: Essays in Feminist Theory.* New York: Crossing Press.

Geden, Oliver, Vivian Scott, and James Palmer. 2018. "Integrating Carbon Dioxide Removal into EU Climate Policy: Prospects for a Paradigm Shift." *Wiley Interdisciplinary Reviews: Climate Change* 9(4): e521.

Gheaus, Anca. 2013. "The Feasibility Constraint on the Concept of Justice." *The Philosophical Quarterly* 63(252): 445–464.

Gilabert, Pablo, and Holly Lawford-Smith. 2012. "Political Feasibility: A Conceptual Exploration." *Political Studies* 60(4): 809–825.

Gössling, Stefan, and Andreas Humpe. 2020. "The Global Scale, Distribution and Growth of Aviation: Implications for Climate Change." *Global Environmental Change* 65: 102194.

Green, Fergus, and Ajay Gambhir. 2020. "Transitional Assistance Policies for Just, Equitable and Smooth Low-Carbon Transitions: Who, What and How?" *Climate Policy* 20(8): 902–921.

Hardy, R. Dean, Richard A. Milligan, and Nik Heynen. 2017. "Racial Coastal Formation: The Environmental Injustice of Colorblind Adaptation Planning for Sea-Level Rise." *Geoforum* 87: 62–72.

Hayashi, A., Sano, F., and Akimoto, K. 2020. "On the Feasibility of Cropland and Forest Area Expansions Required to Achieve Long-Term Temperature Targets." *Sustainability Science* 15(3): 817–834.

Heyward, Jennifer Clare, and Dominic Roser, eds. 2016. *Climate Justice in a Non-Ideal World.* Oxford: Oxford University Press.

Hoekstra, Arjen Y., and Thomas O. Wiedmann. 2014. "Humanity's Unsustainable Environmental Footprint." *Science* 344(6188): 1114–1117.

Honneth, Axel. 1995. *The Struggle for Recognition.* Cambridge: Polity.

Jewell, J., and A. Cherp. 2020. "On the Political Feasibility of Climate Change Mitigation Pathways: Is It Too Late to Keep Warming Below 1.5°C?" *WIREs Climate Change.*

Jones, Phil, Simon Lannon, and Jo Patterson. 2013. "Retrofitting Existing Housing: How Far, How Much?" *Building Research & Information* 41(5): 532–550.

Kidd, Ian James, José Medina, and Gaile Pohlhaus Jr, eds. 2017. *Routledge Handbook of Epistemic Injustice.* New York: Routledge.

Lawford-Smith, Holly. 2013. "Understanding Political Feasibility." *Journal of Political Philosophy* 21(3): 243–259.

Low, Sean, and Stefan Schäfer. 2020. "Is Bio-energy Carbon Capture and Storage (BECCS) Feasible? The Contested Authority of Integrated Assessment Modeling." *Energy Research & Social Science* 60: 101326.

Lyons, Kristen, and Peter Westoby. 2014. "Carbon Colonialism and the New Land Grab: Plantation Forestry in Uganda and Its Livelihood Impacts." *Journal of Rural Studies* 36: 13–21.

Mahony, Martin, and Georgina Endfield. 2018. "Climate and Colonialism." *Wiley Interdisciplinary Reviews: Climate Change* 9(2): e510.

Manne, Kate. 2020. *Entitled: How Male Privilege Hurts Women*. New York: Crown.

Mata, Érika, Joel Wanemark, Vahid M. Nik, and Angela Sasic Kalagasidis. 2019. "Economic Feasibility of Building Retrofitting Mitigation Potentials: Climate Change Uncertainties for Swedish Cities." *Applied Energy* 242: 1022–1035.

McBride, Cillian. 2013. *Recognition*. Cambridge: Polity.

McCauley, Darren, and Raphael Heffron. 2018. "Just Transition: Integrating Climate, Energy and Environmental Justice." *Energy Policy* 119: 1–7.

McIntosh, Peggy. 1988. "White Privilege: Unpacking the Invisible Knapsack." *Peace and Freedom*, July/August.

McIntosh, Peggy. 2012. "Reflections and Future Directions for Privilege Studies." *Journal of Social Issues* 68(1): 194–206.

McTernan, Emily. 2019. "Justice, Feasibility, and Social Science as It Is." *Ethical Theory and Moral Practice* 22(1): 27–40.

Medina, José. 2011. "The Relevance of Credibility Excess in a Proportional View of Epistemic Injustice: Differential Epistemic Authority and the Social Imaginary." *Social Epistemology* 25(1): 15–35.

Miller, David. 2013. *Justice for Earthlings: Essays in Political Philosophy*. Cambridge: Cambridge University Press.

Murphy, Michael W. 2016. "Mapping Environmental Privilege in Rhode Island." *Environmental Justice* 9(5): 159–165.

Nielsen, Kristian S., Paul C. Stern, Thomas Dietz, Jonathan M. Gilligan, Detlef P. van Vuuren, Maria J. Figueroa, Carl Folke, et al. 2020. "Improving Climate Change Mitigation Analysis: A Framework for Examining Feasibility." *One Earth* 3(3): 325–336.

Nightingale, Andrea J. 2017. "Power and Politics in Climate Change Adaptation Efforts: Struggles Over Authority and Recognition in the Context of Political Instability." *Geoforum* 84: 11–20.

Pickering, J., S. Vanderheiden, and S. Miller. 2012. "If Equity's In, We're Out." *Ethics & International Affairs* 26(4): 423–443.

Piggot, Georgia, Michael Boyland, Adrian Down, and Andreea Raluca Torre. 2019. "Realizing a Just and Equitable Transition Away from Fossil Fuels." *Development*: 202033.

Posner, Eric A., and David Weisbach. 2010. *Climate Change Justice*. Princeton: Princeton University Press.

Räikkä, Juha. 1998. "The Feasibility Condition in Political Theory." *Journal of Political Philosophy* 6(1): 27–40.

Rosales, Jon. 2008. "Economic Growth, Climate Change, Biodiversity Loss: Distributive Justice for the Global North and South." *Conservation Biology* 22(6): 1409–1417.

Roser, Dominic. 2015. "Climate Justice in the Straitjacket of Feasibility." In *The Politics of Sustainability: Philosophical Perspectives*, edited by Dieter Birnbacher and May Thorseth, 71–91. New York: Routledge.

Roser, Dominic. 2016. "Reducing Injustice Within the Bounds of Motivation." In *Climate Justice in a Non-Ideal World*, edited by Clare Heyward and Dominic Roser, 83–103. Oxford: Oxford University Press.

Skodvin, Tora, Anne Therese Gullberg, and Stine Aakre. 2010. "Target-Group Influence and Political Feasibility: The Case of Climate Policy Design in Europe." *Journal of European Public Policy* 17(6): 854–873.

Southwood, Nicholas. 2016. "Does 'Ought' Imply 'Feasible'?" *Philosophy & Public Affairs* 44(1): 7–45.

Spencer, Thomas, Michel Colombier, Oliver Sartor, Amit Garg, Vineet Tiwari, Jesse Burton, Tara Caetano, Fergus Green, Fei Teng, and John Wiseman. 2018. "The 1.5 C Target and Coal Sector Transition: At the Limits of Societal Feasibility." *Climate Policy* 18(3): 335–351.

Sullivan, Shannon. 2006. *Revealing Whiteness: The Unconscious Habits of Racial Privilege*. Bloomington: Indiana University Press.

Sullivan, Shannon. 2019. *White Privilege*. Cambridge: Polity Press.

Sullivan, Shannon, and Nancy Tuana, eds. 2007. *Race and Epistemologies of Ignorance*. Albany: SUNY Press.

Taylor, Charles. 1994. *Multiculturalism: Examining the Politics of Recognition*. Princeton: Princeton University Press.

Whitten, Suzanne. 2021. *A Republican Theory of Free Speech: Critical Civility*. New York: Palgrave.

Young, Iris Marion. 2003. "Political Responsibility and Structural Injustice." The Lindley Lecture, University of Kansas, May 5.

Young, Iris Marion. 2009. "Structural Injustice and the Politics of Difference." In *Contemporary Debates in Political Philosophy*, edited by Thomas Christiano and John Christman, 362–383. West Sussex: Wiley-Blackwell.

Chapter 8

Is Climate Justice Feasible? A Psychological Perspective on Challenges and Opportunities for Achieving a Just Climate Regime

Ezra M. Markowitz and Andrew Monroe

Over the past two decades, a significant multidisciplinary body of literature has been developed on the barriers to achieving action on climate change and climate justice (e.g., protection of marginalized populations, fair distribution of the costs for mitigating and adapting to climate change; see special issue in *Climatic Change* co-edited by Grasso and Markowitz 2015). Much of this literature has identified political, sociological, technological, and economic barriers to achieving climate justice goals, many of which are detailed in this volume. Less attention, however, has been paid to the barriers operating at the level of individual or small-group decision-makers, despite broad recognition that climate justice will require significant changes in human behavior at every level of social organization and decision-making, from individuals and households, to local, national, and global communities (Orlove et al. 2020).

Achieving a just future in a climate-changed world will require pursuing political, technological, and social change through pluralistic and democratic public engagement processes. Put another way, a just global response to climate change will not be feasible without deep and meaningful involvement by citizens around the world. In part, this is because climate justice likely demands that some individuals and groups give up some of their privilege in order to meet the needs of many. Yet we know that motivating most individuals to engage with the issue of climate change is an uphill battle (Gifford 2011), particularly at times of great social, political, and economic upheaval. Indeed, one of the great ironies of the climate crisis may be that as the negative impacts of climate change ramp up over the course of the twenty-first century, individuals and communities may become less capable of engaging in meaningful action to combat future consequences (Weber 2006). This will intensify to the extent that climate change increases zero-sum mindsets among the world's richest individuals, communities, and nations. There is, then, a great need to identify strategies to increase public participation in climate justice efforts in the present (Devaney et al. 2020), before such reinforcing feedback loops further inhibit ethical, effective action.

To that end, this chapter aims to help climate advocates, policymakers, and scholars better understand some of the feasibility constraints to achieving just climate action posed by certain psychological factors that shape how people engage with climate change, particularly as a moral issue. We start by identifying some of the key ways in which the nature of the climate crisis can interact with core psychological mechanisms responsible for directing morally relevant behavior. These interactions are thought to inhibit the feasibility of even well-laid plans for a just response to climate change. In the second part of the chapter, we reflect on the question of how (and whether) to account for feasibility when developing a normative vision for a just future given the urgency of the problem at hand. We conclude by presenting a set of evidence-informed strategies that may be helpful in combating psychological barriers to achieving climate justice.

I. PSYCHOLOGICAL BARRIERS TO JUST CLIMATE CHANGE ACTION

Various scholars and practitioners have identified a wide variety of psychological barriers to public engagement and action with respect to climate change (e.g., Gifford 2011; Markowitz and Shariff 2012) as well as potential solutions to address those barriers (e.g., Nielsen et al. 2020). Here, we briefly highlight four psychological mechanisms or barriers that are particularly relevant for scholars interested in addressing issues of climate justice, morality, and feasibility, as each plays a role in shaping people's perceptions of climate impacts and their morally appropriate solutions.

Mismatch between Climate Harms and Moral Intuitions

As has been argued elsewhere (e.g., Markowitz and Shariff 2012; Gardiner 2006), there is a fundamental mismatch between the forms that climate harms take and what drives people's moral intuitions and motivates morally relevant action. It is well-established that moral reactions often drive behavior. However, in the case of climate change, both the harm to the natural environment and the climate change-related harms to human society often fail to fit the prototype for "immoral harm" (i.e., a harm that is extreme or malicious enough to require personal or societal action to redress it). For example, people easily recognize flooding due to an abnormally strong hurricane is harmful, but as a hurricane lacks any malice in its effects, people tend to regard such harms as unfortunate, terrible even, but not immoral. This short circuiting of people's prototypical moral machinery results in several problematic outcomes including less empathy for suffering victims (Čehajić et al. 2009; Decety and Cowell 2015, 2014; Lucas and Kteily 2018), and this is especially extreme if victims are perceived as outgroups. Additionally, the lack of intentional malice in climate-related harms numbs moral emotional reactions that are often necessary to motivate remunerative action (Haidt 2001; Alicke 2000; Ditto and Lopez 1992; Ditto et al. 2009; Clark et al. 2017; Tetlock et al. 2007). This failure is complex and stems from multiple contributing factors.

First, the moral indignation necessary to motivate action requires a clear norm violation (Mikhail 2011; Wheatley and Haidt 2005; Cheng et al. 2013). Problematically, however, climate-relevant behavior lacks clear—or nearly any—behavioral norms for individuals. That is, it isn't immediately obvious what constitutes a norm violation when it comes to individuals' climate-relevant behavior. Is it a violation to live in a large, single-family home? Is it less of a norm violation if that home generates its own electricity via solar panels? What about driving an inefficient vehicle in order to get to one's workplace? Does it matter if there is a reasonable public transit alternative? And what constitutes "reasonable" in the context of behavior that produces greenhouse gas emissions but that simultaneously allows people to meet their basic needs? Indeed, there is not even broad agreement among ethicists who work on climate change regarding what responsibilities, if any, individuals have to pursue or avoid certain actions that may have implications for climate change (e.g., actions that affect one's greenhouse gas emissions; see Sinnott-Armstrong 2005; Nolt 2011).

Critically, in such situations, descriptive norms (i.e., what people believe most others do) powerfully influence behavior and this opens the door for moral slippage. Monroe et al. (2018) demonstrate that people judge the (im)morality of a person's behavior based on how much it deviated from what most other people did, not whether it violated a strict moral principle (e.g., do not steal). Tax evasion, lying to friends, or cheating in a game for real money are judged as relatively acceptable as long as the person doesn't cheat much more than what is believed to be commonplace. In the context of climate behavior where norms for individual behavior are far less defined than norms about cheating and fairness, this suggest that people will accept most climate-damaging behavior as long as they are viewed as close to what is common. Furthermore, Gino et al. (2009) demonstrate that a single blatantly bad actor can substantially shift people's beliefs about moral norms, thereby providing a permission structure for people themselves to engage in worse behavior. Effectively, in the absence of clear, broadly agreed upon norms of behavior, people are motivated to simply avoid being the worst actor of the group. Moreover, as climate failures or corrupt behaviors are widely reported (e.g., companies faking emissions data, deception about recyclables) this may further erode the already ill-defined and permissive moral norms for individual climate behavior.

Second, human morality is ill-equipped to process the scale and impact of climate harm. Quite the opposite, the evolutionary crucible that forged human moral intuitions was designed to respond to situations involving an identifiable victim (typically a single one) and a clear (single) causal agent (Gray et al. 2012). In contrast, climate change is the result of the distributed contributions of multiple (billions of) agents and the harms associated with climate change are unintentional, spread over time, affect many victims, and are sometimes the side effects of behaviors people might conceive of as good (e.g., building a new business, providing for one's children and family).

Climate change represents an existential threat to our species, and its effects are distributed so asymmetrically—disproportionally harming already vulnerable populations—that on its face one would expect these large violations of moral considerations of harm and fairness would intuitively prime anger and outrage (Graham et al. 2009). However, the extreme mismatch between the structure of climate harm and people's moral prototype means that climate harms are, by default, less likely to trigger morally motivating emotions, such as anger and outrage, which likely further blunts popular support for climate justice action. A large body of work demonstrates that anger is a basic motivator of moral action and restoring justice. At a most basic level, anger promotes sensitivity to unfairness (Mullen and Skitka 2006) and is associated with whether people encode events as moral wrongs in need of redress (Gutierrez and Giner-Sorolla 2007; Rozin et al. 1999) versus mere faux pas behaviors that are unfortunate but permissible (Ask and Pina 2011). Furthermore, anger is a key motivator for moral action, linking people's identification of wrong and wrongdoers to specific actions to sanction and punish (Kühne et al. 2015). To the extent that climate change generally fails to make people truly angry, it likely will continue to carry relatively less weight than other pressing issues for which the "ingredients" for moral indignation and outrage are more readily available.

In some ways this makes the problem of climate change unique. In other morally relevant domains people are willing to bear personal burdens in their pursuit of having wrongdoers punished and victims compensated (Fehr and Fischbacher 2004; Fehr et al. 2002; Fehr and Gächter 2002). Moreover, this holds even when the harm doesn't personally affect them. However, this pattern is strongest when harm to a victim is clear, the action was intentional, and there is an identifiable perpetrator. In contrast, when harm is caused by multiple people (either acting together or independently), it weakens corrective moral action. In these cases, people will usually seek to punish the most flagrant responsible offenders, and notably, people wish to punish these individuals substantially less than when an identical harm is caused by a single person (Schein et al. 2020). Similarly, people generally forgive actions that they believe were unintentional or that were not fully under a person's control (Martin and Cushman 2016; Monroe and Malle 2019), and most climate harms fit squarely into these narratives (as Gardiner 2006 has previously argued).

This is not to suggest that people view climate change as a permissible course for the planet. Rather, the lack of clearly defined and broadly accepted norms, the absence of identifiable perpetrators and victims (Schein and Gray 2017; Gray et al. 2012), and the unintentional nature of the harm, all work in concert to mute people's intuitive moral reactions to climate change and hamper morally motivated climate action.

Uncertainty and Moral (dis)Engagement

In addition to lacking clear and identifiable norm violations, violators, and victims, climate change involves deep and pervasive uncertainty on multiple fronts (Working Groups I, II, and III 2015). This further clouds moral judgment and allows individuals and groups to avoid making costly decisions in the present in order to avoid possible harms in the future (Bain et al. 2013; Bal and van den Bos 2012; Monroe et al. 2017); such uncertainty also contributes to the "perfect moral storm" Gardiner laid out nearly fifteen years ago in his seminal work on the topic (Gardiner 2006).

Uncertainty regarding the timing, severity, and distribution of negative impacts is well-demonstrated and has been widely communicated to the public, not only by those skeptical of climate action but by the mainstream climate science community as well (Working Groups I, II, and III 2015). Similarly, there is significant and deep uncertainty about the efficacy of proposed solutions and actions (both at the individual and societal scales), particularly because many solutions that are proposed at the individual level (e.g., reduce one's own carbon footprint) seem woefully inadequate given the scale of the problem. This uncertainty stems in part from the complexity of the physical climate system and from the difficulty in predicting future technological developments; however, the greatest source of uncertainty is human behavior itself, as it remains unclear what we will choose to do as individuals and as a species in the years to come with respect to addressing climate change.

Psychologically, uncertainty is an uncomfortable mental state and one that people generally try to avoid and/or resolve as quickly as possible (Feldman et al. 2019; Kruglanski and Webster 1996). In some cases, salient uncertainties regarding climate change (e.g., Is it happening, what can be done?) can be resolved through learning more about the issue and by taking actions in one's own life to reduce future harms to oneself and others (e.g., homeowners investing in actions that make their homes more resilient to local climate threats). However, much of the uncertainty inherent in climate change is unresolvable, particularly for individuals. As a result, some people address the conflict in less direct and (socially) counterproductive ways, e.g., by disengaging from the issue altogether (Lucas and Davison 2019; Randall 2009).

Although some proponents of climate action dismiss or disparage disengagement as mere "denial" or "sticking one's head in the sand," for many people disengagement with climate change—and particularly moral disengagement—is self-protective and rational (Lertzman 2015). This may be particularly true when individuals feel that there is high uncertainty regarding the efficacy of proposed solutions, as much research within the health and environmental domains demonstrates the importance of perceived efficacy in promoting proactive engagement with threats (Witte 1992). As discussed later, uncertainty about the feasibility of proposed climate justice frameworks or actions may work to further depress citizens' motivation to engage with the issue and, as a result, inhibit support for ameliorative action.

Political Polarization May Inhibit Shared Beliefs about Climate Justice

As has been widely documented elsewhere, climate change has (intentionally) become one of the most politically polarized societal issues over the past two decades (Pew Research Center 2020; McCright and Dunlap 2011; McCright et al. 2014). At the same time, people of different political, religious, and other group identities rely on different core moral principles and values when making judgments about personal and collective responsibility, questions of right and wrong, and the "appropriate" moral frameworks to use in distributing resources across different entities (Graham et al. 2009; Haidt 2007; Koleva et al. 2012). For example, in the United States, it has been shown that liberals and conservatives moralize environmental issues differently (McCright et al. 2014; Feygina et al. 2010; McCright and Dunlap 2011). Additionally, researchers have demonstrated that people's acceptance and use of climate-relevant information is powerfully shaped by their preexisting beliefs and value commitments (Kahan et al. 2012).

Polarization, group-level differences in adherence to particular moral principles, and deep-seated psychological mechanisms that promote biased acceptance and interpretation of new evidence combine to make it challenging to build the consensus and shared understanding of the world necessary to promote collaborative action on climate change. Moreover, this combination of psychological and sociological factors may make it particularly difficult to develop frameworks for climate justice that will be accepted as just and morally right by diverse groups within society, especially within the United States. This may be the case even as public acceptance of climate change as a serious threat facing humanity continues to increase. For example, some proposed frameworks or regimes for just climate policies focus almost exclusively on duties of harm and care, to the exclusion of other values prominent in conservative moral thought (e.g., fealty to one's in-group; Feinberg and Willer 2013; Graham et al. 2009). This may cause people to reflexively oppose climate justice policies or even intentionally flaunt them (as has been the case with responses to mask-mandates in the face of COVID-19). In part, such reactions occur because people tend to believe what they perceive as good or beneficial (e.g., denying climate change earns accolades from my partisans) over what they might otherwise accept as an impartial assessment of the evidence (e.g., climate change is happening) (Cusimano and Lombrozo *in press*). This suggests that feasibility assessments must, then, include considerations not only of what is technologically and politically possible but also considerations of what is shared across disparate groups with respect to fundamental principles upon which moral judgments are based.

Having the Ability to do Something about the Problem Can Inhibit Action

The costs of pursuing climate justice should presumably fall heavily on the most wealthy and powerful individuals, organizations, and governments. Additionally, these individuals and entities are the best positioned to make meaningful strides toward climate justice. However, a large body of social psychological work demonstrates that acquiring power (or wealth) makes people more selfish, more immoral, and more favorable toward the status quo. For example, Maner et al. (2010) showed that obtaining power causes people to behave in more socially inappropriate ways, and Keltner et al. (2003) demonstrated how power causes people to become more reward-focused and less inhibited. Power also promotes more aggressive behavior (Keltner et al. 2003) and less accommodation of others (e.g., decreased politeness in communication settings, e.g., Brown and Levinson 1987). Wealth is similarly associated with more immoral behavior (e.g., Gino and Piersce 2009; Piff et al. 2012).

Because frameworks to promote climate justice will likely require reallocation of both power and wealth (resources) across groups, the pervasive, negative effects of increased power on moral judgment and decision-making represent additional barriers to feasibility in this space, both at the individual and group/organizational levels. Although the negative effects of power and wealth on developing a just climate regime are easily recognizable in global negotiations over climate change, relatively little attention has been paid to the ways in which similar dynamics may play out at the level of individual citizens as they navigate decisions, both large and small, that have implications from a climate perspective. This is an important area for future research.

II. ADDITIONAL CONSIDERATIONS: EFFICACY, CAPITULATION, AND URGENCY

Given the psychological challenges discussed earlier and elsewhere (e.g., Gifford 2011), we see three additional issues at play with respect to the meta-question of whether and to what degree feasibility should be considered when developing proposals for just responses to climate change.

Effects of Feasibility and Efficacy on Moral Judgment and Action

Up to this point, we have used the term "feasibility" in a relatively nonspecific way to refer to the likelihood that some climate action or framework will produce desired outcomes. This loose definition implies both an estimate of the probability that an action or framework will occur or come into force and an estimate of the impact that such action will have in bringing about just climate-related outcomes. Within the social psychological literature, the theoretically well-developed and validated construct of "efficacy" uniquely captures and disentangles both of these dimensions and their impacts on decision-making (e.g., Bandura 1986). Critically, psychologists identify two key features of efficacy relevant to the topic of feasibility, moral judgment, and climate change: (1) perceptions regarding the likelihood that an action (or framework) can be implemented (usually referred to simply as "efficacy"), and (2) perceptions regarding the probable impact that such action will have in bringing about desired outcomes (often referred to as "outcome expectancies" or "response efficacy").

Justice scholars should consider how both components of efficacy likely affect people's motivation to act and their moral judgments of the climate problem itself. Across multiple literatures, there is a clear link between the perceived efficacy of personal actions and group policies and motivation to take ameliorative action to reduce threats (both personal and societal). The role of efficacy in promoting productive responses to threats is most well-established in the health decision-making domain (e.g., Witte 1992), but has also been demonstrated within the context of climate change (e.g., Heath and Gifford 2006; Doherty and Webler 2016). Heath and Gifford (2006), for example, show that people who don't believe their own cooperation with climate-change efforts will make a difference are significantly less likely to take concrete steps to tackle the crisis. In short, if people perceive proposed climate responses to be inefficacious for whatever reason—however morally "just" they may be in theory—one possible consequence could be a decrease in public support for such actions.

Furthermore, perceptions of feasibility and efficacy influence people's beliefs about the nature of the problem itself. If proposed solutions are perceived to be infeasible, one result may be weakened perceptions of climate change action as a moral imperative (possibly for purposes of self-protection; Randall 2009). Although evidence within the climate domain is limited, both experimental and observational results in other domains suggest that people's beliefs about proposed solutions to a societal or personal problem can significantly influence perceptions of the problem itself. For example, Kahan et al. (2012) and others (Campbell and Kay 2014) show that people calibrate their climate-change risk perceptions post-hoc after being told about possible solutions that either align or misalign with their pre-existing commitments (e.g., political ideology); when people learn that an ideologically aligned solution to climate change exists, they are subsequently more likely to perceive the issue as a significant risk.

Together, these dynamics suggest a problematic feedback loop. If people perceive climate change as a problem too daunting to solve or that the efficacy of possible avenues for change are dubious they are likely to become less interested in taking personal action to preserve the climate or to support societal or governmental climate action (Heath and Gifford 2006). That is, if people believe that they cannot (easily) fix the climate they may become disinterested in taking any climate-addressing actions (see also Norgaard 2011). Moreover, these same self-protecting motives change what people perceive to be threatening at all. Essentially, if people believe climate solutions are not feasible, they are likely to recast climate change as less of a threat and thus a lower priority for both personal and collective action.

Explicit Focus on Feasibility Could Backfire

That being said, an overt and overly cautious focus on feasibility—particularly political feasibility—may smack of moral capitulation to some people, particularly those who support strong societal responses to the climate crisis. Similarly, the need to build diverse coalitions of interest groups and policymakers to overcome deeply vested interests in the climate change domain carries the risk of resulting in "feasible" but largely symbolic policy responses to the challenge. If people perceive proposals for climate justice as feasible only because they largely leave the status quo undisturbed, this too may diminish public support for ameliorative action. The literature on sacred values and decision-making (e.g., Tetlock 2003) would seem to further suggest that feasibility that comes at the expense of integrity and real-world impact is unlikely to promote greater public engagement and support for action. To the extent this dynamic is present in the climate change context, it may caution against tying climate justice proposals too closely to requirements of feasibility for feasibility's sake.

Balancing Moral Ideals and Urgency

Finally, the question of whether and how to balance concerns regarding the urgency of the climate crisis with adherence to moral ideals is a potentially important one from the perspective of moral judgment and decision-making. Urgency—perceived or real—can influence decision-making and judgment in both positive and negative ways. In the context of climate change, urgency has been shown to enhance motivation to take action in some cases (e.g., Rooney-Varga et al. 2018) while leading to suboptimal decision-making in others (see Orlove et al. 2020 for an in-depth discussion of urgency and climate change–relevant decision-making). Research in moral judgment and decision-making has similarly shown that time constraints (which promote a sense of urgency) can shift the moral principles through which individuals judge a situation and/or a possible course of action (e.g., Bjorklund 2003). For moral theorists grappling with climate change, these findings suggest a need to carefully examine how urgency may influence the selection or identification of moral ideals being applied when assessing the justness of specific policy proposals and/or actions; that is, the urgency of the climate crisis itself may be an important factor in shaping which moral ideals are seen as relevant when crafting a just response.

III. OVERCOMING PSYCHOLOGICAL BARRIERS TO CLIMATE JUSTICE AND FEASIBILITY CONSIDERATIONS

Although daunting, all of the challenges and key considerations identified earlier are susceptible to disruption through evidence-informed interventions that aim to reframe how individual citizens engage with the underlying issues of climate change and justice. Here, we briefly highlight a subset of approaches identified in recent years that may be useful in either addressing or sidestepping the various challenges discussed earlier. We focus on interventions and tools that have been empirically studied and supported. Broader speculation on additional approaches that may be useful can be found elsewhere (e.g., Markowitz and Shariff 2012).

Activating Moral Machinery for Behavior Regulation

As noted earlier, climate harms are often portrayed as diffuse, physically distant, and temporally extended. This mutes people's moral and emotional responses to the issue. Therefore, talking about climate change in terms of specific, concrete events and the harm they cause (e.g., Hurricane Harvey hitting Houston and Puerto Rico in 2017) is more likely to activate moral reactions and change public opinion. For example, a 2018 poll in North Carolina following the impact of Hurricane Florence showed a 24-point increase in Republicans' belief that global warming was "very likely" to negatively impact North Carolina compared to attitudes documented in 2017 (Elon University 2018).

Although negative reactions and further polarization might result from making recent climate change impacts salient (e.g., explicitly linking recent, specific extreme weather events to climate change; see Chapman and Lickel 2016), doing so may also increase many people's moral motivation to take action. As discussed earlier, work by Gray et al. (2012) suggests that moral outrage (i.e., anger and condemnation) requires clear norm and moral violations in order to become activated, and when people become morally outraged, ameliorative action is much more likely to occur (Gutierrez and Giner-Sorolla 2007; Kühne et al. 2015). Similarly, linking specific events with specific victims can increase empathy for people harmed by climate change (e.g., through the identifiable victim effect; Erlandsson et al. 2015) while possibly reducing compassion fade (i.e., the reduced concern for suffering as the number of sufferers increases) and other psychological mechanisms that people engage in (largely unconsciously) to protect themselves from having to take personally costly action (e.g., emotional down-regulation, Cameron and Payne 2011). Other research suggests that pointing to specific victims may motivate the search for perpetrators (Gray et al. 2014), further increasing motivation for justice and action.

Circumventing Up-Front Costs and Uncertainty about the Future

If one key barrier to implementing climate-justice policies and frameworks is uncertainty—about the impacts of climate change, about the distribution of costs and rewards of action across time and groups—then techniques that help people act in the face of uncertainty may be critical. One technique is to identify concrete and certain benefits of just action. In line with research on construal level theory and decision-making under uncertainty (Trope and Liberman 2003), climate justice policies and efforts should be couched in concrete rather than abstract terms, particularly when discussing the benefits of action either to present or future generations (Trope and Liberman 2003; Wade-Benzoni and Tost 2009); this may feel counterintuitive to advocates more used to couching calls for action in high-minded moral principles. At the same time, proposals that spread the costs of action out across time are generally easier for people to accept (Yoeli et al. 2017), even if future costs are uncertain and the delay in action violates moral ideals.

In a related vein, research has shown that temporal discounting can be short-circuited through various techniques that help reduce the perceived "psychological distance" between the present and the future. For example, Zaval et al. (2015) show that asking people to reflect on how they want to be remembered after they die increased concern about climate change and willingness to donate to environmental nonprofit organizations. More broadly, efforts to frame climate change and justice as components of intergenerational reciprocity and beneficence may be productive in terms of generating public support for costly, upfront action to protect future others (Wade-Benzoni and Tost 2009; Wade-Benzoni et al. 2010).

Communicating uncertainty carefully and purposefully is also critical. Although many climate advocates and communicators appear to be wary of openly discussing uncertainty (and for good reason, Oreskes and Conway 2011), concealing known uncertainties is likely to do more harm than good and should be avoided both for ethical and practical reasons (Markowitz and Guckian 2018). Moreover, recent research has shown that clear, honest communication of uncertainty regarding future impacts of climate change can actually increase citizens' trust in climate scientists and their support for ameliorative policies, though bombarding people with information about possible future impacts of climate change may also overwhelm their ability to process and cope productively with the full extent of uncertainty that exists in this space (e.g., Howe et al. 2019).

Overcoming Political Division

As discussed previously, liberals and conservatives in the United States construct "morality" based on different sets of moral values (Graham et al. 2009), such as harm and care, fairness, loyalty, respect, and sanctity. Although there are group-based differences in the relative weighting of these various core values or principles, most people broadly share many of these values. Yet much messaging around climate change and justice focus on an unnecessarily narrow subset of core moral values, i.e., harm reduction and fairness. Work by Feinberg and Willer (2013) and others suggests that reframing climate justice as being well-aligned with the broader set of moral values that many Americans hold deeply, including the values of sanctity and respect, can promote more politically diverse support for climate action (including greater engagement among ideological conservatives). Similarly, even when talking about climate justice, advocates should be careful about the moral language that they use to describe and frame the problem. Moralizing language can indeed be highly motivating of action when the audience already views the issue in moral terms; however, moralizing language can have opposite, counterproductive effects when audiences don't already view the issue as being a morally relevant one (which research suggests is largely the case for Americans with respect to climate change, particularly but not exclusively for conservatives; Markowitz 2012; Leiserowitz et al. 2016).

Leveraging Power to Promote Care

Overcoming the negative effects of power on moral decision-making is also important for achieving just solutions to the climate crisis, given the dominant role that powerful individuals and nations will play in developing any future climate regimes at the global and national levels. One way to fight the narrowing effects of power is to try to prompt moral expansionism, that is, the feeling that one's moral circle of concern includes a larger group of people (Reed and Aquino 2003; Waytz et al. 2016). A large and robust body of evidence shows that people give preferential treatment to their (moral) in-group members (Crimston et al. 2016; Bastian et al. 2011; Tajfel 1970; Monroe and Plant 2019), and that expanding people's belief about who is in their in-group could reduce some of the selfish biases that come with power (Singer 1981; Laham 2009; Reed and Aquino 2003). Moreover, moral expansiveness may not be limited to extending concern toward typical outgroups; some recent work suggest that people may feel moral obligation toward living and nonliving natural entities (e.g., soil, animals, oceans) as well (Bratanova et al. 2012).

At the same time, entirely overcoming the influence of power on human psychology may be too high a bar to set, particularly under conditions of uncertainty and socio-political discord already being experienced within and beyond the United States. An alternative is to attempt to bypass the influence of power by reframing behavior. Specifically, two types of reframing may be effective. First, reframing climate action as an exceptional, morally praiseworthy behavior rather than a sacrifice may increase compliance with climate action. Research in the context of behavioral responses to the COVID-19 pandemic (Van Bavel et al. 2020), for example, suggests that this sort of reframing can be effective at increasing people's willingness to social distance and wear protective masks (e.g., when you wear your mask, you're protecting and saving others). (For more on the relationship between just climate action and COVID-19, please see Meyer, L. and Araujo, M. in this volume.) Second, reframing climate action in terms of powerful individuals' material self-interest, specifically in terms of leaving a legacy, has the potential to promote personally costly but socially beneficial action (Vandenbergh and Raimi 2015).

IV. CONCLUDING THOUGHTS

If we are to successfully and justly confront the climate crisis, we must collectively act with urgency despite great uncertainty—uncertainty about the true scale of future impacts on human and non-human life, uncertainty about the efficacy of whatever solutions we pursue, and uncertainty about the political, social, economic, and technological pathways we will take over the coming decades (and beyond). In light of this uncertainty, considerations of feasibility with respect to climate justice are clearly important, not only for policymakers and advocates but for ethical theorists as well; from a human decision-making perspective, feasibility is clearly one important factor (among others) that influences whether and how people engage with climate change. Feasibility interacts with the human moral judgment system to shape how people think and feel about the climate crisis in moral terms, which subsequently affects whether and to what extent they choose to respond in productive versus counterproductive ways. Theorists and practitioners alike are thus likely to be more successful in their efforts to bring about a just future if they consider the interconnected ways in which considerations of feasibility shape human moral judgment with respect to this critical societal challenge.

REFERENCES

Alicke, Mark D. 2000. "Culpable Control and the Psychology of Blame." *Psychological Bulletin* 126(4): 556–574.

Ask, Karl, and Afroditi Pina. 2011. "On Being Angry and Punitive: How Anger Alters Perception of Criminal Intent." *Social Psychological and Personality Science* 2(5): 494–499.

Bain, Paul G., Matthew J. Hornsey, Renata Bongiorno, Yoshihisa Kashima, and Daniel Crimston. 2013. "Collective Futures How Projections About the Future of Society Are Related to Actions and Attitudes Supporting Social Change." *Personality and Social Psychology Bulletin* 39(4): 523–539.

Bal, M., and K. van den Bos. 2012. "Blaming for a Better Future: Future Orientation and Associated Intolerance of Personal Uncertainty Lead to Harsher Reactions toward Innocent Victims." *Personality and Social Psychology Bulletin* 38(7): 835–844.

Bandura, Albert. 1986. "The Explanatory and Predictive Scope of Self-Efficacy Theory." *Journal of Social and Clinical Psychology* 4(3): 359–373.

Bastian, Brock, Simon M. Laham, Sam Wilson, Nick Haslam, and Peter Koval. 2011. "Blaming, Praising, and Protecting Our Humanity: The Implications of Everyday Dehumanization for Judgments of Moral Status." *British Journal of Social Psychology* 50(3): 469–483.

Bjorkland, Fredrik. 2003. "Differences in the Justification of Choices in Moral Dilemmas: Effects of Gender, Time Pressure and Dilemma Seriousness." *Scandinavian Journal of Psychology* 44(5): 459–466.

Bratanova, Boyka, Steve Loughnan, and Birgitta Gatersleben. 2012. "The Moral Circle as a Common Motivational Cause of Cross-Situational pro-Environmentalism: Moral Circle and Pro-Environmentalism." *European Journal of Social Psychology* 42(5): 539–545.

Brown, Penelope, and Stephen C. Levinson. 1987. *Politeness: Some Universals in Language Usage (Studies in Interactional Sociolinguistics 4)*. Cambridge: Cambridge University Press.

Cameron, C. D., and B. K. Payne. 2011. "Escaping Affect: How Motivated Emotion Regulation Creates Insensitivity to Mass Suffering." *Journal of Personality and Social Psychology* 100: 1–15.

Campbell, T. H., and A. C. Kay. 2014. "Solution Aversion: On the Relation Between Ideology and Motivated Disbelief." *Journal of Personality and Social Psychology* 107: 809–824.

Čehajić, Sabina, Rupert Brown, and Roberto González. 2009. "What Do I Care? Perceived Ingroup Responsibility and Dehumanization as Predictors of Empathy Felt for the Victim Group." *Group Processes and Intergroup Relations* 12(6): 715–729.

Chapman, Daniel, and Brian Lickel. 2016. "Climate Change and Disasters How Framing Affects Justifications for Giving or Withholding Aid to Disaster Victims." *Social Psychological and Personality Science* 7(1): 13–20.

Cheng, Justin S., Victor C. Ottati, and Erika Price. 2013. "The Arousal Model of Moral Condemnation." *Journal of Experimental Social Psychology* 49(6): 1012–1018.

Clark, Cory J., Roy F. Baumeister, and Peter H. Ditto. 2017. "Making Punishment Palatable: Belief in Free Will Alleviates Punitive Distress." *Consciousness and Cognition* 51(Supplement C): 193–211.

Crimston, Charlie R., Paul G. Bain, Matthew J. Hornsey, and Brock Bastian. 2016. "Moral Expansiveness: Examining Variability in the Extension of the Moral World." *Journal of Personality and Social Psychology* 111(4): 636–653.

Cusimano, C., and T. Lombrozo. 2021. "Morality Justifies Motivated Reasoning in the Folk Ethics of Belief." *Cognition*. Online: 104513.

Decety, Jean, and Jason M. Cowell. 2014. "The Complex Relation between Morality and Empathy." *Trends in Cognitive Sciences* 18(7): 337–339.

Decety, Jean, and Jason M. Cowell. 2015. "Empathy, Justice, and Moral Behavior." *AJOB Neuroscience* 6(3): 3–14.

Devaney, Laura, Pat Brereton, Diarmuid Torney, Martha Coleman, Constantine Boussalis, and Travis G. Coan. 2020. "Environmental Literacy and Deliberative

Democracy: A Content Analysis of Written Submissions to the Irish Citizens' Assembly on Climate Change." *Climatic Change* 162: 1965–1984.

Ditto, Peter H., David A. Pizarro, and David Tannenbaum. 2009. "Motivated Moral Reasoning." In *Moral Judgment and Decision Making*, edited by Daniel M. Bartels, Christopher W. Bauman, Linda J. Skitka, and Douglas L. Medin, 307–338. San Diego: Elsevier Academic Press.

Ditto, Peter H., and David F. Lopez. 1992. "Motivated Skepticism: Use of Differential Decision Criteria for Preferred and Nonpreferred Conclusions." *Journal of Personality and Social Psychology* 63(4): 568–584.

Doherty, Kathryn, and Thomas Webler. 2016. "Social Norms and Efficacy Beliefs Drive the Alarmed Segment's Public-Sphere Climate Actions." *Nature Climate Change* 6: 879–884.

Elon University. 2018. "The Impact of Hurricane Florence on North Carolina Voters."

Erlandsson, Arvid, Fredrik Björklund, and Martin Bäckström. 2015. "Emotional Reactions, Perceived Impact and Perceived Responsibility Mediate the Identifiable Victim Effect, Proportion Dominance Effect and In-Group Effect Respectively." *Organizational Behavior and Human Decision Processes* 127: 1–14.

Fehr, Ernst, and Simon Gächter. 2002. "Altruistic Punishment in Humans." *Nature* 415(6868): 137–140.

Fehr, Ernst, and Urs Fischbacher. 2004. "Third-Party Punishment and Social Norms." *Evolution and Human Behavior* 25(2): 63–87.

Fehr, Ernst, Urs Fischbacher, and Simon Gächter. 2002. "Strong Reciprocity, Human Cooperation, and the Enforcement of Social Norms." *Human Nature* 13(1): 1–25.

Feinberg, Matthew, and Robb Willer. 2013. "The Moral Roots of Environmental Attitudes." *Psychological Science* 24(1): 56–62.

FeldmanHall, Oriel, and Amitai Shenhav. 2019. "Resolving Uncertainty in a Social World." *Nature Human Behaviour* 3(5): 426–435.

Feygina, Irina, John T. Jost, and Rachel E. Goldsmith. 2010. "System Justification, the Denial of Global Warming, and the Possibility of 'System-Sanctioned Change.'" *Personality and Social Psychology Bulletin* 36: 326–338.

Gardiner, Stephen M. 2006. "A Perfect Moral Storm: Climate Change, Intergenerational Ethics and the Problem of Moral Corruption." *Environmental Values* 15: 397–413.

Gifford, Robert. 2011. "The Dragons of Inaction: Psychological Barriers that Limit Climate Change Mitigation and Adaptation." *American Psychologist* 66(4): 290.

Gino, Francesca, and Lamar Pierce. 2009. "The Abundance Effect: Unethical Behavior in the Presence of Wealth." *Organizational Behavior and Human Decision Processes* 109(2): 142–155.

Gino, Francesca, Shahar Ayal, and Dan Ariely. 2009. "Contagion and Differentiation in Unethical Behavior: The Effect of One Bad Apple on the Barrel." *Psychological Science* 20(3): 393–398.

Graham, Jesse, Jonathan Haidt, and Brian A. Nosek. 2009. "Liberals and Conservatives Rely on Different Sets of Moral Foundations." *Journal of Personality and Social Psychology* 96(5): 1029–1046.

Grasso, Marco, and Ezra M. Markowitz. 2015. "The Moral Complexity of Climate Change and the Need for a Multidisciplinary Perspective on Climate Ethics." *Climatic Change* 130: 327–334.

Gray, Kurt, Chelsea Schein, and Adrian F. Ward. 2014. "The Myth of Harmless Wrongs in Moral Cognition: Automatic Dyadic Completion from Sin to Suffering." *Journal of Experimental Psychology: General* 143(4): 1600.

Gray, Kurt, Liane Young, and Adam Waytz. 2012. "Mind Perception Is the Essence of Morality." *Psychological Inquiry* 23(2): 101–124.

Gutierrez, Roberto, and Roger Giner-Sorolla. 2007. "Anger, Disgust, and Presumption of Harm as Reactions to Taboo-Breaking Behaviors." *Emotion* 7(4): 853–868.

Haidt, Jonathan. 2001. "The Emotional Dog and Its Rational Tail: A Social Intuitionist Approach to Moral Judgment." *Psychological Review* 108(4): 814–834.

Haidt, Jonathan. 2007. "The New Synthesis in Moral Psychology." *Science* 316(5827): 998–1002.

Heath, Yuko, and Robert Gifford. 2006. "Free-Market Ideology and Environmental Degradation: The Case of Belief in Global Climate Change." *Environment & Behavior* 38: 48–71.

Howe, Lauren C., Bo MacInnis, Jon A. Krosnick, Ezra M. Markowitz, and Robert Socolow. 2019. "Acknowledging Uncertainty Impacts Public Acceptance of Climate Scientists' Predictions." *Nature Climate Change* 9: 863–867.

Kahan, Dan M., Ellen Peters, Maggie Wittlin, Paul Slovic, Lisa L. Ouellette, Donald Braman, and Gregory Mandel. 2012. "The Polarizing Impact of Science Literacy and Numeracy on Perceived Climate Change Risks." *Nature Climate Change* 2: 732–735.

Keltner, Dacher, Deborah H. Gruenfeld, and Cameron Anderson. 2003. "Power, Approach, and Inhibition." *Psychological Review* 110(2): 265–284.

Koleva, Spassena P., Jesse Graham, Ravi Iyer, Peter H. Ditto, and Jonathan Haidt. 2012. "Tracing the Threads: How Five Moral Concerns (Especially Purity) Help Explain Culture War Attitudes." *Journal of Research in Personality* 46(2): 184–194.

Kruglanski, Arie W., and Donna M. Webster. 1996. "Motivated Closing of the Mind: 'Seizing' and 'Freezing'." *Psychological Review* 103: 263–281.

Kühne, Rinaldo, Patrick Weber, and Katharina Sommer. 2015. "Beyond Cognitive Framing Processes: Anger Mediates the Effects of Responsibility Framing on the Preference for Punitive Measures." *Journal of Communication* 65(2): 259–279.

Laham, Simon M. 2009. "Expanding the Moral Circle: Inclusion and Exclusion Mindsets and the Circle of Moral Regard." *Journal of Experimental Social Psychology* 45(1): 250–253.

Leiserowitz, Anthony, Edward Maibach, Connie Roser-Renouf, Geoff Feinberg, and Seth Rosenthal. 2016. *Climate Change in the American Mind*. New Haven, CT: Yale Program on Climate Change Communication, Yale University and George Mason University.

Lertzman, Renee. 2015. "In Climate Change, Psychology Often Gets Lost in Translation." *Climate Psychology Alliance* 24.

Lucas, Brian J., and Nour S. Kteily. 2018. "(Anti-)Egalitarianism Differentially Predicts Empathy for Members of Advantaged versus Disadvantaged Groups." *Journal of Personality and Social Psychology* 114(5): 665–692.

Lucas, Chloe H., and Aidan Davison. 2019. "Not 'Getting on the Bandwagon': When Climate Change is a Matter of Unconcern." *Environment and Planning E: Nature and Space* 2(1): 129–149.

Maner, Jon K., Michael P. Kaschak, and John L. Jones. 2010. "Social Power and the Advent of Action." *Social Cognition* 28(1): 122–132.

Markowitz, Ezra M. 2012. "Is Climate Change an Ethical Issue? Exploring Young Adults' Beliefs About Climate and Morality." *Climatic Change* 114: 479–495.

Markowitz, Ezra M., and Azim F. Shariff. 2012. "Climate Change and Moral Judgment." *Nature Climate Change* 2: 243–247.

Markowitz, Ezra M., and Meghan Guckian. 2018. "Climate Change Communication: Challenges, Insights and Opportunities." In *Psychology and Climate Change*, edited by Susan Clayton and Christie Manning, 35–63. San Diego, CA: Elsevier.

Martin, Justin W., and Fiery Cushman. 2016. "Why We Forgive What Can't Be Controlled." *Cognition* 147(February): 133–143.

McCright, Aaron M., and Riley E. Dunlap. 2011. "The Politicization of Climate Change and Polarization in the American Public's Views of Global Warming, 2001–2010." *The Sociological Quarterly* 52(2): 155–194.

McCright, Aaron M., Riley E. Dunlap, and Chenyang Xiao. 2014. "Increasing Influence of Party Identification on Perceived Scientific Agreement and Support for Government Action on Climate Change in the United States, 2006–12." *Weather, Climate, and Society* 6(2): 194–201.

Mikhail, John M. 2011. *Elements of Moral Cognition: Rawls' Linguistic Analogy and the Cognitive Science of Moral and Legal Judgment*. New York, NY: Cambridge University Press.

Monroe, Andrew E., and Bertram F. Malle. 2019. "Testing a Socially-Regulated Blame Hypothesis: People Systematically Update Moral Judgments of Blame." *Journal of Personality and Social Psychology* 116(2): 215–236.

Monroe, Andrew E., and E. Ashby Plant. 2019. "The Dark Side of Morality: Prioritizing Sanctity over Care Motivates Denial of Mind and Prejudice toward Sexual Outgroups." *Journal of Experimental Psychology: General* 148(2): 342–360.

Monroe, Andrew E., Kyle D. Dillon, Steve Guglielmo, and Roy F. Baumeister. 2018. "It's Not What You Do, But What Everyone Else Does: On the Role of Descriptive Norms and Subjectivism in Moral Judgment." *Journal of Experimental Social Psychology* 77: 1–10.

Monroe, Andrew E., Sarah E. Ainsworth, Kathleen D. Vohs, and Roy F. Baumeister. 2017. "Fearing the Future? Future-Oriented Thought Produces Aversion to Risky Investments, Trust, and Immorality." *Social Cognition* 35(1): 66–78.

Mullen, Elizabeth, and Linda J. Skitka. 2006. "Exploring the Psychological Underpinnings of the Moral Mandate Effect: Motivated Reasoning, Group Differentiation, or Anger?" *Journal of Personality and Social Psychology* 90(4): 629–643.

Nielsen, Kristian S., Sander van der Linden, and Paul C. Stern. 2020. "How Behavioral Interventions Can Reduce the Climate Impact of Energy Use." *Joule* 4: 1613–1616.

Nolt, John. 2011. "How Harmful Are the Average American's Greenhouse Gas Emissions?" *Ethics Policy and Environment* 14(1): 3–10. doi: 10.1080/21550085.2011.561584.

Norgaard, Kari. 2011. *Living in Denial: Climate Change, Emotions, and Everyday Life*. Cambridge, MA: MIT Press.

Oreskes, Naomi, and Erik M. Conway. 2011. *Merchants of Doubt: How a Handful of Scientists Obscured the Truth on Issues from Tobacco Smoke to Global Warming*. USA: Bloomsbury Publishing.

Orlove, Ben, Rachael Shwom, Ezra Markowitz, and So-Min Cheong. 2020. "Climate Decision-Making." *Annual Review of Environment and Resources* 45: 271–303.

Pew Research Center. 2020. "Americans See Spread of Disease as Top International Threat, Along With Terrorism, Nuclear Weapons, Cyberattacks." 1–23.

Piff, P. K., D. M. Stancato, S. Cote, R. Mendoza-Denton, and D. Keltner. 2012. "Higher Social Class Predicts Increased Unethical Behavior." *Proceedings of the National Academy of Sciences USA* 109(11): 4086–4091.

Randall, R. 2009. "Loss and Climate Change: The Cost of Parallel Narratives." *Ecopsychology* 1(3): 118–129.

Reed, Americus, and Karl F. Aquino. 2003. "Moral Identity and the Expanding Circle of Moral Regard toward Out-Groups." *Journal of Personality and Social Psychology* 84(6): 1270–1286.

Rooney-Varga, J. N., J. D. Sterman, E. Fracassi, T. Franck, F. Kapmeier, V. Kurker, E. Johnston, A. P. Jones, and K. Rath. 2018. "Combining Role-Play with Interactive Simulation to Motivate Informed Climate Action: Evidence from the World Climate Simulation." *PLoS One* 13(8): e0202877.

Rozin, P., L. Lowery, S. Imada, and J. Haidt. 1999. "The CAD Triad Hypothesis: A Mapping between Three Moral Emotions (Contempt, Anger, Disgust) and Three Moral Codes (Community, Autonomy, Divinity)." *Journal of Personality and Social Psychology* 76(4): 574–586.

Schein, Chelsea, Joshua Conrad Jackson, Teresa Frasca, and Kurt Gray. 2020. "Praise-Many, Blame-Fewer: A Common (and Successful) Strategy for Attributing Responsibility in Groups." *Journal of Experimental Psychology: General* 149(5): 855–869.

Schein, Chelsea, and Kurt Gray. 2017. "The Theory of Dyadic Morality: Reinventing Moral Judgment by Redefining Harm." *Personality and Social Psychology Review*, May. 1088868317698288.

Singer, Peter. 1981. *The Expanding Circle*. Oxford: Clarendon Press.

Sinnott-Armstrong, Walter. 2005. "It's Not My Fault: Global Warming and Individual Moral Obligations." In *Perspectives on Climate Change: Science, Economics, Politics, Ethics*, edited by Walter Sinnott-Armstrong and Richard B. Howarth, 258–308. Oxford: Elsevier.

Tajfel, Henri. 1970. "Experiments in Intergroup Discrimination." *Scientific American* 223(5): 96–102.

Tetlock, Philip E. 2003. "Thinking About the Unthinkable: Coping With Secular Encroachments on Sacred Values." *Trends in Cognitive Science* 7: 320–324.

Tetlock, Philip E., Penny S. Visser, Ramadhar Singh, Mark Polifroni, Amanda Scott, Sara Beth Elson, Philip Mazzocco, and Phillip Rescober. 2007. "People as Intuitive Prosecutors: The Impact of Social-Control Goals on Attributions of Responsibility." *Journal of Experimental Social Psychology* 43(2): 195–209.

Trope, Yaacov, and Nira Liberman. 2003. "Temporal Construal." *Psychological Review* 110(3): 403–421.

Van Bavel, Jay, Katherine Baicker, Paulo S. Boggio, Valerio Capraro, Aleksandra Cichocka, Mina Cikara, Molly J. Crockett, et al. 2020. "Using Social and Behavioural Science to Support COVID-19 Pandemic Response." *Nature Human Behaviour* 4(5): 460–471.

Vandenbergh, Michael P., and Kaitlin T. Raimi. 2015. "Climate Change: Leveraging Legacy." *Ecology Law Quarterly* 42(1): 139.

Wade-Benzoni, Kimberly A., Harris Sondak, and Adam D. Galinsky. 2010. "Leaving a Legacy: Intergenerational Allocations of Benefits and Burdens." *Business Ethics Quarterly* 20(1): 7–34.

Wade-Benzoni, Kimberly A., and Leigh Plunkett Tost. 2009. "The Egoism and Altruism of Intergenerational Behavior." *Personality and Social Psychology Review* 13(3): 165–193.

Waytz, Adam, Ravi Iyer, Liane Young, and Jesse Graham. 2016. "Ideological Differences in the Expanse of Empathy." In *Social Psychology of Political Polarization*, edited by Piercarlo Valdesolo and Jesse Graham, 61–77. New York, NY: Taylor & Francis.

Weber, E.U. 2006. "Experience-Based and Description-Based Perceptions of Long-Term Risk: Why Global Warming Does Not Scare Us (Yet)." *Climatic Change* 77(1–2): 103–120.

Wheatley, Thalia, and Jonathan Haidt. 2005. "Hypnotic Disgust Makes Moral Judgments More Severe." *Psychological Science* 16(10): 780–784.

Witte, Kim. 1992. "Putting the Fear Back Into Fear Appeals: The Extended Parallel Process Model." *Communications Monographs* 59(4): 329–349.

Working Groups I, II, and III to the Fifth Assessment Report of the Intergovernmental Panel on Climate Change. 2015. *Climate Change 2014: Synthesis Report*. Geneva, CH: IPCC.

Yoeli, Erez, David Budescu Amanda Carrico, Magali Delmas, J. R. DeShazo, Paul Ferraro, Hale Forester, Howard Kunreuther, Rick Larrick, Mark Lubell, Ezra M. Markowitz, Bruce Tonn, Michael Vandenbergh, and Elke U. Weber. 2017. "Behavioral Science Tools for Energy and Environmental Policy." *Behavioral Science & Policy Journal* 3: 69–80.

Zaval, Lisa, Ezra M. Markowitz, and Elke U. Weber. 2015. "How Will I Be Remembered? Conserving the Environment for the Sake of One's Legacy." *Psychological Science* 26(2): 231–236.

Chapter 9

Climate Change, Individual Preferences, and Procrastination

Fausto Corvino

The scientific community almost unanimously agrees that it is imperative to reach the target of net-zero carbon emissions by 2050 to avoid causing irreversible changes to the climate, with all the negative consequences that would follow for both human and non-human animals (IPCC 2018, chap. 2). Yet, the political and individual response to this looming threat continues to lack the needed momentum (Keating 2020; Hodgson 2019). This failure is in stark contrast to the responsiveness that contemporary societies show in the face of other threats, including environmental ones. For example, in the faces of floods, earthquakes, or similar events, people are willing to change their working, production, and consumption habits, and to support supererogatory redistribution of wealth to deal with collective threats. The most recent example of this is the response to the COVID-19 pandemic. In only a few days, almost half the world's population radically changed their day-to-day routines in order to contain the pandemic (See Araujo and Meyer, L. in this volume for more on this topic.)

Some believe that it is epistemic obstacles that make it difficult to motivate people to take concrete action to address climate change (CC), because unlike nuclear or pandemic threats, for example, most of the negative consequences of CC are not knowable by everyone through direct experience but require a high degree of trust in the work and forecasts of experts (Almassi 2012), and this makes the game easy for those who want to deny or reduce the scope of the phenomenon both for specific interests and for mere moral self-absolution (see, e.g., Gelfert 2013). Others argue that it is difficult for people to see CC as a threat that will affect them, or at least as a threat for which they carry some responsibility (Jamieson 1991; Gardiner 2011). This is because CC is a complicated phenomenon that unfolds over time, and its negative effects are predictable only within ranges of risk percentages. Still others contend that it is difficult to feel a strong moral motivation to address CC. This is because its spatial and temporal scope gives the impression that it is possible to delay action indefinitely. After all, why should a person living in Vienna reduce her meat consumption if the main victims of CC are located in distant areas subject to flooding or desertification? Moreover, why should she do her part if richer and more powerful people do not? Isn't it the responsibility of institutions to deal with CC, rather than individual citizens (Jamieson 2010; Peeters et al. 2019; Persson 2017)?

My theoretical assumption in this chapter is that both the epistemic and the moral obstacles have been largely overcome by the recent waves of climate mobilization, by the serious commitment made by the media on the subject, and not least from the recent and frequent climate-related environmental tragedies (such as the latest floods in Venezia; see BBC 2019). According to extensive studies by the Pew Research Center, over the last decade there has been an increase in the perception of CC as a major threat facing human beings. More specifically, CC is now perceived as the most dangerous threat—more than terrorist attacks or nuclear wars—by respondents in thirteen of the twenty-six countries surveyed, representing countries on five continents (Fagan and Huang 2019; Poushter and Huang 2019). Another recent study by the Yale Program on Climate Change Communication suggests that almost 70 percent of Americans are "somewhat worried" by CC, while half of them report having personally felt the negative consequences of CC (Leiserowitz et al. 2019). Similarly, individual motivation to bear a share of the CC mitigation costs has increased considerably in recent years. A study published by the Energy Policy Institute and the AP-NORC Center for Public Affairs Research at the University of Chicago (2019) found that more than two-thirds of Americans would be willing to pay a carbon tax if they were assured that the proceeds would be used for natural environment restoration. While another study revealed that over 60 percent of British people would be willing to pay to substantially reduce the number of deaths—estimated to be 7,000 additional deaths from 2050 and 12,000 from 2080 onward—in the country associated with CC in the years to follow (Graham et al. 2019).

Yet, despite the fact that people today are much more convinced of the looming and dangerous nature of CC, so much so that they declare themselves willing to make sacrifices to contain it, the motivational gap seems to persist. This poses a serious problem of feasibility with respect to CC mitigation policies, which require not only political dynamism (from both representatives and voters) but also the direct willingness of individuals to shoulder a small part of the mitigation burden (Tan 2015). The focus of this chapter will be on the latter. In particular, I will identify the psychological phenomena that encourage individual procrastination in relation to the implementation of personal CC mitigation strategies. These phenomena can properly be categorized as feasibility hurdles to acting on our accepted moral commitments with regard to CC since their presence makes it less likely that we will act to reduce our individual contribution to CC. Thus, it is important that we understand both how these problems arise and how we can dismantle them as obstacles that stand in the way of personal change.

My central argument in this chapter is that the main reason for people's continued lack of responsiveness to CC can be explained by a "short circuit" that leads them to decisions that do not reveal their true preferences regarding CC. A preference is "true" if it is one that an individual forms when she looks at intertemporal choices from a position that is temporally neutral. Such a position is one that allows her to form rational judgments about the maximization of her individual well-being, that is, judgments that consider a possible discount rate of future utility but that are nonetheless immune from the psychological distortions which, once the moment of choice is approaching, can influence her to sacrifice medium- or long-term benefits (whose utility is duly discounted) to obtain fewer benefits in the immediate future. I claim that this short circuit is caused by two psychological phenomena. First, as the possibility of early rewards gets closer, individuals discount further future utility and reverse their preferences. Second, individuals struggle to order their preferences in a transitive manner when there is an opportunity to obtain small marginal utilities (or avoid disutilities) in the short-term.[1] If we can understand how these phenomena manifest in the context of personal decision-making about CC mitigation, then we can devise solutions that are well-suited to reduce their influence. Toward this end, I offer three strategies that will help individuals resist the temptation to procrastinate on this front. I conclude that while the mechanisms that lead to procrastination can decrease the feasibility of personal action in the face of CC, properly understanding and implementing plans to overcome these phenomenon will render personal efforts aimed at CC mitigation much more feasible.

I. INTERTEMPORAL PREFERENCE REVERSAL

Let's start with the psychological phenomenon whereby people tend to assign more and more value to immediate utility as the moment of decision approaches, often resulting in a complete reversal of their temporally neutral preferences. In this section I will explain why this continuous preference reversal can trap the individual in an infinite loop of procrastination, in which she never carries out the choice she considers the best one, at least, as we will see, from a temporally neutral perspective.

The intertemporal dimension of CC is usually treated by both economists and philosophers in terms of social discount rates, which is the rate at which we must discount costs and benefits realized at equal intervals in the future (Rendall 2019). Let's say that we have to decide how to invest a given amount of money, M, in a project that will be financed over the period of t to $t1$, but will provide returns (in terms of utilities) that extend to the later time $t2$. Many people value enjoying those utilities earlier, rather than later, and how much they do so sets a discount rate. This means they would hold that a marginal increase (or decrease) of utility U that occurs at time $t1$ is more valuable than an equal increase (or decrease) of U that occurs at $t2$. Put another way, if they could invest $2 at $t1$ to obtain $2.2 at $t2$, the existence of a discount rate could imply that their investment actually reduces U, because the value of $2.2 at $t2$ is lower than the value of $2 at t.

There are two basic normative reasons that justify the social discount rate. The first one is that when people can choose whether to enjoy a given benefit sooner or later, they prefer to have it sooner, and this should be considered when we evaluate individual well-being. The second reason is that we commonly assume that economic growth rates are sustained over time, so the principle of diminishing marginal utility makes it mandatory that benefits accrued to a poor and to a rich society are not treated equally (see also Nordhaus 2007). Although many people criticize the practice of discounting on normative grounds (see, e.g., Caney 2014; Tarsney 2017), the social discount rate is not inconsistent with individual preferences; indeed, it is supposed to be an indication of how members of a given social group on average assess the utility achieved in the future. Accordingly, once we agree on a given social discount rate, someone who conforms to it in her cost-benefit analysis is supposed to stick to her "true" preferences, that is to say, stick to those preferences that maximize her well-being in light of the fact that the value of utility diminishes over time.

A different phenomenon that is often confused with the discounting of future utility is what Manuel Utset (2010, 256) calls "the immediacy multiplier." Consider the following case. At time t John buys a new car that is worth 100 coins, and the market value of this car is expected to decrease by 2 coins every year for the first two years, by 10 coins per year at the end of the third year, by 20 coins per year at the end of the fourth year, and by 25 coins per year from the fifth year onward until its market value drops to zero. Accordingly, at t John decides that the strategy that would maximize his well-being is to exchange his car within $t1$, that is within the first two years from purchase. Indeed, John is aware of the fact that the value of his utility decreases over time, let us suppose at a rate of 4 percent per year. Thus, if at the end of the second year he waits one year more to change his car, the marginal value of his loss is not 10 utils but rather 9.6 utils.[2] But the rational strategy remains: bear the cost of changing the car before entering the third year, that is, before the annual devaluation rate multiplies fivefold.

So, we can take this as John's preference expressed at time *t* to be his true preference.

However, as soon as the end of the second year approaches (*t1*), chances are that John begins to look at the investment costs of exchanging his car (say 4 utils) vis-à-vis the loss of entering into the third year in a different way than he did at *t*. It may happen that at *t1* he applies an "immediacy multiplier" to the investment cost of 4 utils, such that at *t1* he considers the disvalue of the investment costs to be greater now than he did at *t*. Let us imagine, for example, that John values a benefit obtained right now three times more than a benefit obtained a year later; hence his immediacy multiplier is 3. So, once *t1* arrives, the perceived disvalue of the investment cost for exchanging car is no longer 4 utils, as it was when John was still at *t*, but rather 12 utils (4×3). When the immediacy multiplier is sufficiently high, as in this case, a preference reversal may occur (Andreou 2007a, 185–187, 2007b, 236–237). Specifically, at *t* exchanging the car within the second year is valued more highly than waiting until the third year, but at *t2*, the opposite is true.[3]

The problem with intertemporal preference reversal is that it can give rise to a chain of procrastination. If John follows the strategy that, at *t1*, "seems" to maximize his well-being and accordingly waits until the third year to change cars, then at time *t2* (the end of the third year) he again ends up in the same situation. At *t1* he is determined to change his car at *t2* so as to avoid losing 20 more utils. Yet, when *t2* comes, the investment costs that John should pay for avoiding entering the fourth year, 14 utils [2 utils (first year's depreciation) + 2 utils (second year's depreciation) + 10 utils (third year's depreciation)] gets multiplied by the rate of preference for immediate consumption, that is 3; hence John is faced with a choice between renouncing 42 utils (14×3) now and renouncing 32.64 utils [34 utils (the sum of annual depreciations from the first to the fourth year)[4]—1.36 utils (the 4 percent of 34, that is, the yearly discount rate)] at the end of next year, which can give rise to another preference reversal. This behavior is likely to be repeated several times until John ends up in the situation that was the worst for him at *t*, that is, the one in which his car no longer has market value.

The problem with intertemporal preference reversal is that it can lead an individual into a "procrastination loop" (Andreou 2007a, 189) in which he never reveals in choice his true preferences. In the case of John, his true preferences are the ones he formed at *t* about *t1* (or at *t1* about *t2*, and so forth). The question at this point, however, is why we should consider John's preferences as true at *t* but not at *t1*. Moreover, why do we consider it to be rational to apply a discount rate at *t*, but irrational to apply the immediacy multiplier at *t1*? At *t* John is making an intertemporal evaluation of utility, in which he takes into consideration the pros and cons of obtaining well-being sooner or later, without what Utset defines as "transient distortions brought about by the prospect of experiencing immediate gratification" (2010, 256). In other words, at t John understands that the best strategy for maximizing his well-being is to spend either 2 coins at the end of the first year or 4 coins at the end of the second year, while also considering the opportunity cost of investing money to safeguard the value of the car and the fact that within a few years his income could be greater. Yet, the psychological distortions that arise at the moment in which John is called to open his wallet lead him to abandon this strategy. Another way to understand why the preference that John matures at t about what he has to do at $t1$ is less true than the preference that he matures at $t1$ about what he has to do at $t1$, is to consider that the second preference is purely temporary. It develops only with the approach of $t1$ and if performed it gives rise to remorse immediately after $t1$: that is, if John does not make the investment at $t1$, he regrets it right afterward, because he understands that this delay would prevent him from maximizing his well-being in the medium run. Hence he is committed to do the second best thing, that is, changing car at $t2$. Yet, at $t2$ John will form a different preference than he had at $t1$ about $t2$, and this is why he is locked into in a procrastination loop that progressively makes him worse off than he could have been if he stuck to his true preferences initially.

I suggest that intertemporal preference reversal is one of the two main explanations behind general procrastination, and I believe that there are some intrinsic features of CC that lead to this form of procrastination (see also Andreou 2007b, 240–242). As in the car example, CC mitigation is a matter of incurring costs now to avoid incurring ever-increasing costs in the future. And when it comes to making spending decisions, the incentive to postpone them is quite high because the negative consequences of doing so (increased CC effects + higher costs of mitigation) will not be felt until later (Persson 2017). The intertemporal separation of costs and benefits renders CC more similar to paradigmatic situations of procrastination due to preference reversal, as in the car case, than to other global threats, such as terroristic attacks, nuclear wars, and epidemics. What characterizes these latter phenomena is that only one part of the costs can be postponed, namely those related to investments for containment, while the negative effects of this postponement cannot be postponed, namely the risk of attacks or the deaths due to the epidemics. This helps to explain why these other threats usually receive proportionate responses, both in terms of public investments and in terms of changes in individual lifestyles, while CC does not.

II. INTRANSITIVE PREFERENCES

The second characteristic of CC that leads to procrastination is the inability of many individuals to acknowledge the transitivity of preferences with respect to mitigation strategies. Preferences have transitive properties in the following way: if an individual considers option X superior to option Y, and option Y superior to option Z, then she should also consider X superior to Z. The inability to acknowledge this characteristic is due to the fact that CC is the sum of countless individual emissions that taken separately are negligible (Sinnott-Armstrong 2005; Jamieson 2010).[5] This is a challenge that is often discussed in relation to assigning individual moral responsibility for CC. Why should I consider it morally wrong to travel by car rather than by bus if the emissions caused by a one-day drive are neither sufficient nor necessary to cause CC? This same concern is also related to the question of whether individual moral responsibility for CC is political—understood a responsibility to engage in political advocacy for the adoption of appropriate climate policies—or also interpersonal.

Yet, a less evident, but equally important, result of the structure of CC is that it provides the psychological prerequisites for a continuous postponement of changes in individual behaviors that are necessary for mitigation. Because of the small disutility of a marginal postponement and the huge disutility of an immediate change in individual routine, the rational thing to do is to continuously postpone the change, until the individual ends up in a situation in which she prefers not having postponed from the first moment. For even those individuals—the majority, I guess—who care not only about themselves but also about their children and grandchildren, or more generally about their community, know that the negative effect of an additional steak or car trip on the group or on the people they care about is much less than the disutility they would personally have to endure from a day of abstinence (from meat or from their car). In this sense, the individual gets trapped in a procrastination loop in which she does not reverse her preferences, as in the previous examples, but rather she cannot maintain the transitivity of her preferences over time.

Let's consider, for example, a person who finds out he has high cholesterol and has to stop eating his favorite food, cheese. What is the right moment to stop eating cheese, or in other words, what is the best strategy that maximizes individual well-being in the face of both the risks stemming from high cholesterol and the great pleasure spawned by every single slice of cheese? At time t our cheese-enthusiast is still far from suffering the negative effects that high cholesterol might have on his body, and one more slice of cheese might give him significant pleasure without altering his cholesterol level. Thus, he might have good reason to prefer eating one more slice. But a second slice of cheese would also have no effect on his cholesterol level, so he might prefer eating two more slices rather than just one. If we take N to be the level of cheese consumption of this person at t, and W to be the minimum amount of cheese slices which, summed to N, can give rise to severe health problems, we might order the preferences of the cheese-enthusiast in the following way: $N+1 < N+2 < N+3 < N+4 < N+5 < \ldots < N+W < N+1$ (See also Andreou 2007a, 187–188; Thaler and Tversky 1992; Aldred 2007).

As we can see, the structure of the "cheese problem," which is the result of a long series of consumption actions that outside the sum do not have any impact on the health of our friend, prevent him from ordering his preferences in a transitive way. While the cheese-enthusiast has reasons to prefer any single act of procrastination, this chain leads him to a level of cheese consumption that he considers worse than not having procrastinated from the first moment he discovered his risks associated with high cholesterol, namely t. This sort of intransitivity of individual preferences is a different phenomenon from the intertemporal preference reversal, since at no time in the cheese case does the individual reverse his preferences. The latter remain stable and intransitive. Yet, both phenomena give rise to similar procrastination loops. More specifically, the intransitivity of individual preferences in cases as the one just described creates a sort of individual tragedy of the commons. For the rational thing to do at each subsequent interval is to keep on exploiting the "individual" pool of resources,[6] which in this case is the body's ability to absorb cheese without causing a dangerous accumulation of cholesterol. Yet, the pursuit of maximum individual utility risks progressively dragging the individual in a way that actually minimizes his utility.

Something similar happens with individual actions that can mitigate CC, as for example going by bus instead of car or investing money in the thermal insulation of a home, with the further complication that the problem of procrastination is exacerbated by the possibility of freeriding on others. For in the cheese case, and more generally in classic cases of procrastination that are discussed in the psychological literature, it is the same individual who procrastinates who also suffers the consequences of marginal delays and there is no way to delegate to others the actions that one continues to postpone. Conversely, even the individual who is strongly committed to changing her behaviors to mitigate CC knows that a failure to meet her commitment will have an insignificant effect on CC. On the other hand, in a situation in which all other individuals do not abandon the status quo, the committed individual knows that her change of behavior has an insignificant effect on CC.

Accordingly, the right way to look at the procrastination of individual actions that could mitigate CC is as the procrastination of single actions that, in order to achieve the objective that motivated them, must be accompanied by a series of similar actions carried out by other individuals over whom the procrastinator has no control. More simply, while the cheese-enthusiast can experience both the positive effects of fulfilling his commitment (better blood analysis) and the negative effects of procrastination (the risk of raising his cholesterol too high), the CC mitigator cannot touch either the positive or the negative effects of his change of transport (for instance). His commitment has to be strong enough that he sticks to it simply because it is collectively rational to do so: that is, if everyone renounces the maximization of her well-being (by renouncing freeriding, for example), it is possible for everyone to avoid ending up in the worst situation, a climate out of control.

III. OVERCOMING CLIMATE CHANGE PROCRASTINATION

Procrastination is the last step in the chain of obstacles that limit the feasibility of personal CC mitigation. It arises when an individual has understood the dangers of CC, is convinced that she wants to play her part toward remedying the collective problem (Baatz 2014), but nonetheless continues to postpone the actions that she considers morally just. Accordingly, I maintain that the proper way to look at the feasibility problem of CC with respect to what individuals could do is not in terms of a motivational gap (Roser 2016) or of moral corruption (Gardiner 2011), but rather in terms of the implementation of already established motivations. Being distinct with regard to the diagnosis of the feasibility obstacles that hinder CC mitigation will therefore allow us to be distinct with regard to the solution around these obstacles.

If we are to make individual mitigation efforts feasible, we must first acknowledge that the two types of procrastination described earlier require two different strategies of containment because their causes are different. The second thing to note is that not every procrastinator is aware of the psychological mechanisms that are at stake in his constant postponement of the things he feels are important. In this respect, I think that, following Utset (2010, 257), it is useful to distinguish between naïve, sophisticated, and partially naïve procrastinators. In the case of intertemporal preference reversal, a naïve procrastinator is convinced, at time t, that at time $t1$ he will be able to carry out what he deemed rational at time t; a sophisticated procrastinator is aware that when the time comes to collect a benefit or postpone a loss, she will reverse the preferences she ordered at time t; and a partially naïve procrastinator is aware that a sort of utility multiplier can increase the value of an early reward (or loss) but she does not foresee that this phenomenon can reverse individual preferences. In the case of stable and intransitive preferences, a naïve procrastinator is not aware of the circularity of her preferences; a sophisticated procrastinator knows that at some point she will consider not having procrastinated from the first moment as preferable to the situation she ends up in through marginal delays that, taken separately, are rational; and a partially naïve procrastinator knows that she might have incentives to postpone the tasks she is committed to, yet she ignores that this might lead her into a procrastination loop.

The third thing that it is important to highlight if we are to overcome procrastination on CC mitigation is that many strategies and suggestions for overcoming classic cases of procrastination have been formulated (see for example Steel 2010; Lamia 2017). It is impossible to consider this enormous body of information in the space that remains in this chapter. What I shall do, instead, is to focus on those strategies that I consider more effective with respect to CC. As we shall see, there are three broad classes of actions a procrastinator can undertake to try to stick to her true preferences. However, the precondition for a procrastinator to overcome procrastination is that she understands being a procrastinator; that is, she must understand that there is a problem with the rationality of her judgments. Accordingly, the strategies for softening procrastination are accessible only to sophisticated procrastinators. On the other hand, what we can do with both naïve and partially naïve procrastinators is to make them aware of the risk they face when their choices don't actually reveal their preferences, as explained in the section "Intertemporal Preference Reversal."

Having made these points, the three broad classes of actions in response to procrastination that a self-conscious procrastinator can undertake are the following: (i) pre-committing to the costly action that is likely to be postponed at a later stage; (ii) raising the costs of marginal postponements; or (iii) reducing the disutility of the action that is likely to be postponed. Below, I analyze each of these strategies in turn.

Pre-Commitment

Let us go back to the car case. Suppose that John, the car buyer, is a sophisticated procrastinator. At t, he prefers to exchange his car by $t1$, but he also knows that when $t1$ comes he will prefer waiting one more year, and after one year he will prefer to wait one year more. Accordingly, John makes this request to the car dealer at the moment of purchase: "Please, let's make a contract, according to which in twenty-four months I give the car back, and you deduct its value from the purchase of a new car, which I now pledge to purchase." This sort of pre-commitment can neutralize intertemporal preference reversal and can allow John to reveal in choice the preferences he has at time t (See also Utset 2010, 257; Elster 2000, 1–87; Verbeek 2007).

Pre-commitment can be useful with respect to the feasibility of those individual investments that could reduce emissions (and so contribute to CC mitigation), of which individuals perceive both the moral urgency and the economic convenience in the medium term, but which risk being continually postponed due to a preference reversal that occurs as the time of payment approaches. Some examples can be the purchase of low-energy appliances, the thermal insulation of homes, the installation of solar panels, and so on. Let us suppose that after having fixed the car affair, John realizes that installing solar panels on the roof of his house would be both morally good and utility maximizing for him in the medium run. It would allow him to stick to his moral commitment to do his part with respect to CC and at the same time he knows that his investment will be returned in full in just five years, at which time he will start saving money on his electricity bills. Yet, when the time comes to sign the check to the solar panel dealer, the disutility of withdrawing the amount due for the panels from his bank account increases so much because of the utility multiplier, that it ends up outweighing both the utility of reducing his emissions in the years to come and the savings he will earn on his bills after five years.

A solution to this problem could consist in making the spending decision binding at time t, while leaving the actual spending at time $t1$. As in the car case, we can imagine, for example, that at time t John asks the panel dealer for a contract, in which it is established that at $t1$ (e.g., one year later) he will buy the solar panel. By doing this, John makes sure that at the time he decides whether or not to make the medium-run investment, he values costs and benefit in a "rational" way, that is by balancing early costs and future benefits (which include both the utility stemming from sticking to his environmental commitment and the utility of paying lower bills). Through this pre-commitment, John can discount future utility without falling prey to the immediacy multiplier. In other words, it helps John to evaluate his investment from a psychologically neutral position, and this allows him to implement his true preferences.

Pre-commitment, when available, works quite well with intertemporal preference reversal. If sellers of all sorts of eco-friendly products—from household appliances to solar panels to electric cars—offered purchasing plans of this sort, they would greatly help consumers implement the decisions they consider to be morally right (and in some cases also economically worthwhile) with respect to CC. And in doing so they would help to make the transition to less-polluting consumption items more feasible, in those cases in which the consumer has already matured a preference about a reduction in her individual emissions but she is unable to reveal this preference in her choices because she is trapped in a procrastination loop. Obviously, when the time of spending comes, the consumer might regret having signed the contract because of the immediacy multiplier, but at that point she would have no choice but to stick to her preferences, assuming that the costs of breaching the contract are sufficiently high.

It might be argued that after the consumer regrets having signed the first pre-commitment contract she will not sign another contract of this kind. This is not necessarily true, for the idea is that pre-commitment helps the individual to realize her true preferences, helping her to overcome the momentary alteration of rationality that occurs at the time of payment. Accordingly, when this obstacle has been overcome, we can expect the individual to start enjoying the benefits of the investment vis-à-vis the costs. Certainly, it may happen that someone like John may realize it is not worth investing a lot of money to help mitigate CC, because, for example, it is more important to invest in his daughter's education, or that the expected savings on the bill are not actually substantial once the maintenance costs of the panels are factored in. Yet, these contingencies have nothing to do with intertemporal preference reversal, but correspond instead to a change in preferences, which is beyond the scope of this chapter.

Raising the Costs of Marginal Postponements

We'll now consider the other form of procrastination, the one that is caused by intransitive preferences. How can the cheese lover get off the loop that leads him to high cholesterol? One simple solution is to recommend this person to renounce maximizing his utility. At the moment he receives his first blood test he has to stop eating cheese and in so doing he does not run the risk of losing control of his cholesterol. Yet, this is in contradiction with individual rationality, to the extent that it does not maximize well-being so l long as the risk is the result of many small actions rather than a single event. So why should our friend abstain from one more slice of cheese before quitting altogether? This is not simply a pedantic academic objection, but a very practical problem. Recommending that someone renounce an action that can augment his well-being without harming anyone else is hardly a winning strategy. For the individual incentive to violate the recommendation with respect to the single action remains quite high, regardless of the risks that could be encountered in the medium or in the long run.

Instead, we can try to render any single postponement inconvenient (see also Andreou 2010, 209–210). This would reconcile the strategy of risk containment with individual rationality. Let us imagine that, like John, the cheese lover is also a sophisticated procrastinator, hence he is aware of the fact that if he keeps on consuming cheese he will end up in a procrastination loop. Accordingly, he asks his partner to hide the cookies he likes so much for breakfast, every time he eats a slice of cheese. This agreement would substantially increase the disutility of the marginal consumption of cheese, which is no longer the small impact on cholesterol levels but also includes having to forego cookies for breakfast. Under these circumstances, the rational thing to do is to renounce cheese from the first blood test, that is, before entering the procrastination loop. As in the car case, the trick consists in leading the individual to decide in advance upon a given course of action by raising the "costs" before the loop begins.

The implications of this simple model for increasing the feasibility of individual actions aimed at mitigating CC are quite relevant. Just think of all those changes in consumption and transportation habits that can help reduce an individual's impact on CC: using public transport instead of a car, eliminating meat from one's diet, sorting waste properly, and making the use of electricity more efficient inside one's home, to name a few. As with the cheese case, these are situations in which the disutility of marginal postponements are negligible (one more steak or one more trip by car will not change the fortunes of the planet), whereas the disutility of the change in habits is huge (just think about how painful it could be for a meat lover to suddenly become a vegetarian or for a person used to going to work in a car to have to adapt to public transport). Thus, the rational strategy is to keep on postponing the change in habits, until the individual record of emissions gets so high that the individual has fallen short of her commitment to play her part in the containment of CC. In comparison to the cheese case, individual CC mitigation has the further complication that the individual emissions that result from a complete procrastination loop have a minimal, almost imperceptible, effect on the global mathematics of CC. Accordingly, the right way to look at the failure of individual rationality is with respect to moral preferences: the individual *wishes* to play her part in CC mitigation, regardless of the practical consequences of this, but due to continuous postponements she does not do so.

One solution is to render marginal postponements costly. Let us focus, for example, on the commuter who is resolute in abandoning the car in favor of the bus, but keeps on postponing this change in her routine for the reasons just discussed. One way to raise the disutility of one more day by car is to ask someone, her flatmate for example, to add some costs to any single postponement. Suppose that the commuter makes the following request to the flatmate: "Please, every time that you see me taking the car in the morning, leave your dirty dishes in the sink, so when I come back, I have to wash them before I can cook." If the flatmate accepts the task, the commuter would find herself having to balance two different costs every morning. On the one hand, there is the disutility of knowing that she is not living according to her personal moral commitment with respect to CC *plus* the disutility of having to wash someone else's dishes in the evening. On the other hand, there is the disutility of a longer and less comfortable trip by bus instead of car. Chances are that the first sum of disutilities can outweigh the second disutility and hence block the procrastination loop.

As Andreou (2007b, 243–248) has correctly pointed out, however, these external control mechanisms face the further problem of "second-order procrastination." That is, the same reasons that induce the commuter to continuously postpone the change in means of transport might lead her to postpone adopting the solution. After all, imagine that the commuter wakes up in the morning and knows that she will once again delay the desired course of action. Why would she set up a mechanism to prevent her from procrastinating if she wants to procrastinate? Here, too, the trick is to anticipate the solution by separating the time of the decision to adopt a strategy to prevent procrastination and its actual implementation. Imagine our commuter asks her flatmate to implement her sanction in the following week, adding also that if she changes her mind and withdraws the sanction warrant, the flatmate should not take it into account but should rather stick to the first warrant. This would allow the commuter to set in motion a procrastination containment mechanism before the procrastination instinct comes into play, either in the first-order or second-order variants.

Reducing the Disutility of the Action That Is Likely to be Postponed

Another lever that can be used to combat both intertemporal preference reversal and intransitive preferences is to ease the malaise and discomfort that is associated with the course of action that is sought to be postponed. Admittedly, in situations like the solar panel investment, it is quite difficult to find something fun or enjoyable in signing a check. Yet, when we are in the presence of single, small actions that change our routine, it is possible to associate them with small rewards or to try to socialize them so as to render them more enjoyable.

Let us focus on the bus/car case. We have seen that we can raise the costs of marginal postponements through something like the "dishes penalty." But we can also think of lessening the "costs" of prompt action. We could transform the penalty mechanism into a reward mechanism. For example, we can suppose the commuter entrusts her precious bottle of Japanese single malt whiskey to her flatmate and advances the following request: "Please, keep it hidden from me, and let me pour a glass in the evening only on the condition that I took the bus to get to work that morning." Another strategy is to create a new reward, instead of depriving yourself of something and then taking it back as a reward. Our commuter might convince herself that by taking the bus instead of the car for one week she gets enough exercise (walking to and from bus stops) that she can eat dessert when she goes out to dinner with friends (see also Andreou 2010, 211–212). The commuter can lower the disutility of the change in routine and hence reduce her incentive to get into the procrastination loop. A further possibility is to render the change in routine more enjoyable in itself, without any reward. An easy solution can be to focus on the positive aspects of the bus trip that exist but risk being underappreciated:

the chance to read the newspaper on the way, to take a nap, to meet new people, and so on. A more complex solution is to try to render the trip more enjoyable. For example, the commuter might try to persuade her friends or colleagues to take the bus as well, so that they can have a chat together on the way there and back (see Heath and Anderson 2010, 242–244).

These strategies, seemingly trivial, can play a role in the feasibility of those routine changes that are necessary to implement one's moral preferences with respect to CC. This is particularly true with regard to those actions—such as changing means of transport, consuming products with low emission impact, reducing the use of air heating and cooling in the summer and winter—which are perceived by many as just, and lack associated costs (indeed, in many cases they lead to savings), but are nonetheless difficult to implement because of non-economic disutilities (the renunciation of the comfort of the car, the convenience in terms of time and effort of a diet based on meat, the pleasure of living in a very cold environment in summer). In these cases, neither the reward mechanism nor the strategies that aim to make the change in routine more pleasant, eliminate the non-economic disutility associated with the change in routine, nor do they aim to correct their subjective perception over time, but instead aim to introduce new counterbalancing benefits, so as to encourage the individual to do what she considers right.

IV. CONCLUSION

There is a fourth way to avoid procrastination, which I have not considered in this chapter: taxes, subsidies, and state sanctions. Do you want to convince John to invest in solar panels? Just let him deduct the costs from his taxes. Do you want to convince someone to stop using the car? You can raise taxes on fuel, create restricted traffic zones, or increase parking costs. Do you want to encourage the purchase of green appliances? Give subsidies to producers so they can lower costs.

These practices are valid, though if taxation is not well calibrated it risks unfairly allocating the costs of the energy transition. The same applies to subsidies, which could unfairly enrich some people to the detriment of others. And complete prohibitions on some activities are likely to be so radical that it will be hard to get them approved.

Yet, the purpose of this chapter was to explore why many people fall short of their personal moral commitment with respect to CC, thus stalling any attempt to engage in personal CC mitigation. The explanation is that two characteristics of CC expose motivated people to a status quo bias: the temporal dispersions of costs and benefits and the fact that CC is the result of a huge number of actions that taken separately are negligible. Accordingly, I discussed three different strategies to overcome procrastination in personal CC mitigation. The advantage of framing the feasibility of individual duties of climate justice in terms of helping people to reveal their true preferences, instead of assuming that people have morally wrong motivations and focus on how to change them, is that it enhances their moral autonomy. In other words, it leads them to implement what they consider as morally compelling without resorting to coercive practices, thus avoiding, among other things, the costs of enforcement.

NOTES

1. The most insightful contributions on these two topics should be credited to Chrisoula Andreou (2005, 2007a, 2007b, 2010) and Manuel A. Utset (2006, 2010).

2. Here I am assuming, for simplicity, that utility is directly proportional to coins, that is, to purchasing power, hence I consider, for example, 10 coins to be equivalent to 10 utils.

3. In more analytical terms, if we call option A "exchanging the car within the second year," and option B "waiting until the third year," we may say that a preference reversal occurs for John if at t he thinks A>B, but at $t2$ he considers A<B.

4. 34 utils = 2 utils (first year's depreciation) + 2 utils (second year's depreciation) + 10 utils (third year's depreciation) + 20 utils (fourth year's depreciation).

5. John Nolt (2011) convincingly argues that it would be wrong to consider as negligible the harm caused by the annual emissions of an average American (that is the amount of emissions you obtain by dividing the sum of American emissions by the American population). Yet, in this chapter I am a focusing on the negligible impact, in terms of emissions, of a marginal routine activity. This is compatible with the claim that the effects of a huge number of these activities repeated throughout the year are by no means negligible.

6. Conversely, in classic collective tragedies of the commons we are concerned with "common pool resources" (see Schlager 2002).

REFERENCES

Aldred, Jonathan. 2007. "Intransitivity and Vague Preferences." *Journal of Ethics* 11(4): 377–403.

Almassi, Ben. 2012. "Climate Change, Epistemic Trust, and Expert Trustworthiness." *Ethics and the Environment* 17(2): 29–49.

Andreou, Chrisoula. 2005. "Going from Bad (Or Not so Bad) to Worse: On Harmful Addictions and Habits." *American Philosophical Quarterly* 42(4): 323–331.

Andreou, Chrisoula. 2007a. "Understanding Procrastination." *Journal for the Theory of Social Behaviour* 37(2): 183–193.

Andreou, Chrisoula. 2007b. "Environmental Preservation and Second-Order Procrastination." *Philosophy and Public Affairs* 35(3): 233–248.

Andreou, Chrisoula. 2010. "Coping with Procrastination." In *The Thief of Time: Philosophical Essays on Procrastination*, edited by Chrisoula Andreou and Mark D. White, 206–215. Oxford: Oxford University Press.

Baatz, Christian. 2014. "Climate Change and Individual Duties to Reduce GHG Emissions." *Ethics, Policy and Environment* 17(1): 1–19.

BBC. 2019. "Venice Floods: Climate Change Behind Highest Tide in 50 Years, Says Mayor." November 13. https://www.bbc.com/news/world-europe-50401308.

Caney, Simon. 2014. "Climate Change, Intergenerational Equity and the Social Discount Rate." *Politics, Philosophy and Economics* 13(4): 320–342.

Elster, Jon. 2000. *Ulysses Unbound: Studies in Rationality, Precommitment, and Constraints*. Cambridge: Cambridge University Press.

Energy Policy Institute and the AP-NORC Center for Public Affairs Research at the University of Chicago. 2019. "Is the Public Willing to Pay to Help Fix Climate Change? – Findings From A November 2018 Survey of Adults Age 18 and Older." http://www.apnorc.org/projects/Documents/EPIC%20fact%20sheet_v4_DTP.pdf.

Fagan, Moira, and Christine Huang. 2019. "A Look at How People Around the World View Climate Change." *Pew Research Center*, April 18. https://www.pewresearch.org/fact-tank/2019/04/18/a-look-at-how-people-around-the-world-view-climate-change/.

Gardiner, Stephen M. 2011. *A Perfect Moral Storm: The Ethical Tragedy of Climate Change*. Oxford: Oxford University Press.

Gelfert, Axel. 2013. "Climate Scepticism, Epistemic Dissonance, and the Ethics of Uncertainty." *Philosophy and Public Issues (New Series)* 3(1): 167–208.

Graham, Hilary, Siân de Bell, Nick D. Hanley, Stuart W. Jarvis, and Piran C. L. White. 2019. "Willingness to Pay for Policies to Reduce Future Deaths from Climate Change: Evidence from a British Survey." *Public Health* 174: 110–117.

Heath, Joseph, and Joel Anderson. 2010. "Procrastination and the Extended Will." In *The Thief of Time: Philosophical Essays on Procrastination*, edited by Chrisoula Andreou and Mark D. White, 233–252. Oxford: Oxford University Press.

Hodgson, Camilla. 2019. "UK Set to Miss Goal to Cut Carbon Emissions to 'Net Zero' by 2050." *Financial Times*, October 11. https://www.ft.com/content/6fb7fce0-ec37-11e9-a240-3b065ef5fc55.

IPCC (Intergovernmental Panel on Climate Change). 2018. *Global Warming of 1.5°C. An IPCC Special Report on the Impacts of Global Warming of 1.5°C Above Pre-industrial Levels and Related Global Greenhouse Gas Emission Pathways, in the Context of Strengthening the Global Response to the Threat of Climate Change, Sustainable Development, and Efforts to Eradicate Poverty*, edited by Valérie Masson-Delmotte, Panmao Zhai, Hans-Otto Pörtner, Debra Roberts, Jim Skea, Priyadarshi R. Shukla, Anna Pirani, et al. In Press. Accessed October 2, 2020. https://www.ipcc.ch/sr15/.

Jamieson, Dale. 1991. "The Epistemology of Climate Change: Some Morals for Managers." *Society & Natural Resources* 4(4): 319–329.

Jamieson, Dale. 2010. "Climate Change, Responsibility, and Justice." *Science and Engineering Ethics* 16(3): 431–445.

Keating, Dave. 2020. "France and Germany Miss EU Climate Plan Deadline." *Forbes*, January 23. https://www.forbes.com/sites/davekeating/2020/01/23/france-and-germany-miss-eu-climate-plan-deadline/#1e3351b511b0.

Lamia, Mary. 2017. *What Motivates Getting Things Done: Procrastination, Emotions, and Success*. Lanham, Maryland: Rowman & Littlefield Publishers.

Leiserowitz, Anthony, Edward Maibach, Seth Rosenthal, John Kotcher, Matthew Ballew, Matthew Goldberg, and Abel Gustafson. 2019. "Climate Change in the American Mind: December 2018, Executive Summary." *Yale Program on Climate Change Communication*, January 22. https://climatecommunication.yale.edu/publications/climate-change-in-the-american-mind-december-2018/2/.

Nolt, John. 2011. "How Harmful Are the Average American's Greenhouse Gas Emissions?" *Ethics, Policy and Environment* 14(1): 3–10.

Nordhaus, William D. 2007. "A Review of the Stern Review on the Economics of Climate Change." *Journal of Economic Literature* 45(3): 686–702.

Peeters, Wouter, Lisa Diependaele, and Sigrid Sterckx. 2019. "Moral Disengagement and the Motivational Gap in Climate Change." *Ethical Theory and Moral Practice* 22(2): 425–447.

Persson, Ingmar. 2017. "Climate Change-The Hardest Moral Challenge?" *Res Publica* 8(1–2): 3–13.

Poushter, Jacob, and Christine Huang. 2019. "Climate Change Still Seen as the Top Global Threat, but Cyberattacks a Rising Concern." *Pew Research Center*, February 10. https://www.pewresearch.org/global/2019/02/10/climate-change-still-seen-as-the-top-global-threat-but-cyberattacks-a-rising-concern/.

Rendall, Matthew. 2019. "Discounting, Climate Change, and the Ecological Fallacy." *Ethics* 129(3): 441–463.

Roser, Dominic. 2016. "Reducing Injustice Within the Bounds of Motivation." In *Climate Justice in a Non-Ideal World*, edited by Clare Heyward and Dominic Roser, 83–103. Oxford: Oxford University Press.

Schlager, Edella. 2002. "Rationality, Cooperation, and Common Pool Resources." *American Behavioral Scientist* 45(5): 801–819.

Sinnott-Armstrong, Walter. 2005. "It's Not My Fault: Global Warming and Individual Moral Obligations." In *Perspectives on Climate Change: Science, Economics, Politics, Ethics*, edited by Walter Sinnott-Armstrong and Richard B. Howarth, 221–252. Amsterdam: Elsevier.

Steel, Piers. 2010. *The Procrastination Equation: How to Stop Putting Things Off and Start Getting Stuff Done*. New York: Harper Perennial.

Tan, Kok-Chor. 2015. "Individual Duties of Climate Justice Under Non-Ideal Conditions." In *Climate Change and Justice*, edited by Jeremy Moss, 129–147. Cambridge: Cambridge University Press.

Tarsney, Christian. 2017. "Does A Discount Rate Measure The Costs Of Climate Change?" *Economics and Philosophy* 33(3): 337–365.

Thaler, Richard H., and Amos Tversky. 1992. "Preference Reversal." In *The Winner's Curse: Paradoxes and Anomalies of Economic Life*, edited by Richard H. Thaler. New York: The Free Press.

Utset, Manuel A. 2006. "When Good People Do Bad Things: Time-Inconsistent Misconduct & Criminal Law." *FSU College of Law, Public Law Research Paper No. 232.* SSRN: https://ssrn.com/abstract=895734.

Utset, Manuel A. 2010. "Procrastination and the Law." In *The Thief of Time: Philosophical Essays on Procrastination*, edited by Chrisoula Andreou and Mark D. White, 253–273. Oxford: Oxford University Press.

Verbeek, Bruno. 2007. "Rational Self-Commitment." In *Rationality and Commitment*, edited by Fabienne Peter and Hans Bernhard Schmidt, 150–174. Oxford: Oxford University Press.

Chapter 10

COVID Pandemic and Climate Change

An Essay on Soft Constraints and Global Risks

Lukas H. Meyer and Marcelo de Araujo

In this chapter, we examine the measures that have been deployed to address the 2020 COVID-19 pandemic as a model for the assessment of the feasibility of climate goals (CGs). One methodological strategy for the assessment of the feasibility of measures to address climate change consists in examining how human beings, in the face of similar challenges in the past, managed to achieve effective solutions with different degrees of success. As Jessica Jewell and Aleh Cherp put the problem: "The most straightforward way to judge political feasibility is by historic examples" (Jewell and Cherp 2019, 8). One problem with this approach is that humanity has never faced an event akin to dangerous climate change. But recently this situation changed radically. In 2020, humanity bore witness to an event that now lends itself as a strong model for the assessment of the feasibility of CGs: the new coronavirus SARS-CoV-2 pandemic (also known as COVID-19 pandemic, or simply the 2020 pandemic).

The emergence of the new coronavirus brought about a major global crisis, as all states and civil society around the world had to implement drastic measures to limit the spread of a new disease (COVID-19, or simply COVID). These measures, unprecedented in recent world history, resembled the war efforts during World Wars I and II. On March 16, 2020, the French President Emmanuel Macron, for instance, declared: "Nous sommes en guerre" (we are at war) (*Paris Match* 2020). In a press conference on March 22, former President Donald Trump followed suit: "We're at war, in a true sense we're at war" (The White House 2020). In the months that followed, further comparisons with a world war loomed large in the press.

In order to abate the pandemic, states had to intervene in very sensitive and critically important areas of social life concerning, for instance, freedom of movement and assembly, right to privacy and education, as well as the right to run a business and serve customers without imposing on them burdens such as requiring personal information for the purpose of contact tracing, social distancing, or the compulsory use of face masks. During the pandemic, states also had to introduce special rules for access to scarce resources such as food, medicine, and medical care. States likewise forced airlines to ground long-distance flights, which indirectly led to a significant, if temporary, global reduction of CO_2 concentration in the atmosphere (Le Quéré 2020; El Zowalaty et al. 2020).

But, here is the paradox: in spite of mounting evidence that unmitigated climate change will have devastating consequences for humanity over the next decades—consequences, indeed, even more harmful than the pandemic is likely to cause—governments and civil society have been far less engaged in adopting effective measures to avert dangerous climate change. How can we explain this paradox?

One might perhaps suggest that, as of 2020, humanity had more time to mitigate greenhouse emissions, and to adapt to climate change in the future, than it had to contain the advance of the new coronavirus. But, as we intend to show in this chapter, this assumption is misleading. Given the relevant time constraints at play in each threat, immediate action is required, whether to address a pandemic or to avoid dangerous climate change. One might also suggest that the effects of a pandemic are more apparent than the effects of climate change, as the number of deaths resulting from COVID can be followed on a daily basis in real time (see e.g., Worldmeter). However, some of the deadly consequences of climate change have already become quite visible as well. Consider, for instance, the heatwave of 2003 in Europe. It claimed over 70,000 lives (Robine et al. 2008). According to one report published by the World Meteorological Organization in 2019, extreme weather events are expected to occur more frequently over the next decades (WMO 2020; IPCC 2018, 53). Even though the number of statistical premature deaths resulting from climate change is estimated to be much higher than the number of deaths hitherto attributed to COVID (Watts 2019), climate change has not elicited measures comparable to those that have been implemented to address the 2020 pandemic. Thus, the paradox remains.

In order to address this question, we intend to draw a distinction between "hard constraints" and "soft constraints," now common in the philosophical debate on political feasibility (Gilabert and Lawford-Smith 2012; Lawford-Smith 2013; Brennan and Sayre-McCord 2016; Jewell and Cherp 2019). We argue that in order to tackle both pandemics (not only the 2020 pandemic) and dangerous climate change successfully, both mitigation and adaptation measures are called for. Having identified two particularly robust soft constraints that we call "structural constraints" (or geopolitical constraints) and "proximity constraints," we argue that the soft constraints impeding the success of CGs, on the one hand, and on the successful implementation of health policy measures to address pandemics, on the other, have different degrees of malleability. Our analysis proves that the obstacles to averting the climate crisis are much more difficult to overcome. The trans- and intergenerational versions of the structural and proximity constraints may well represent some of the most critical issues human beings have to address in the present, if they intend to bequeath a safe and hospitable environment to the future generations. Yet, we conclude this chapter with a note of optimism for the younger generation, growing up in a post-pandemic world.

I. HARD CONSTRAINTS AND SOFT CONSTRAINTS

After explaining the difference between hard and soft constraints we will introduce and analyze "structural constraints" and "proximity constraints" and show how they hinder the pursuance of adaptation and mitigation policies with respect to both pandemics and the climate change crisis. In the next section we show why these constraints have proved to be more malleable in responding to the pandemic.

Hard and soft constraints differ in kind. If one or more actors have G as a goal, two different kinds of obstacles may stand in their way to achieving G. Some obstacles cannot be overcome by means of social policies, institutional design, or human decision-making, for these obstacles are imbedded in the principles of logic and the laws of nature, or depend on the availability of natural resources. Human beings can, for example, devise social policies to promote the development of carbon sinks (goal G), but the laws of chemistry that underlie phenomena such as photosynthesis or ocean salinity cannot be altered by means of policy-making. The laws of chemistry, thus, represent a "hard constraint" on the feasibility of G. If at least one hard constraint stands in the way of G, there is not much point in asking how difficult it is to achieve G. In this case, G simply cannot be achieved. As far as hard constraints are concerned, G is either feasible or not feasible. Hard constraints, thus, impose a binary (rather than scalar) value on the feasibility of G.

But even if no hard constraint stands in the way of G, "soft constraints" may still represent an obstacle to its achievement. While hard constraints are imbedded in factors such as the laws of nature and the principles of logic, soft constraints arise from social factors and from some aspects of human psychology. Unlike hard constraints, soft constraints are "malleable": actors can succeed in achieving G depending on their capacity to overcome, for example, socio-cultural, economic, legal, or geopolitical constraints, or depending on their capacity to change lifestyles that prevent them from achieving G. Soft constraints impose a scalar value on the feasibility of G. In the absence of hard constraints, some goals, therefore, are more feasible than others.

In calling a constraint "soft," we are not suggesting that it can be easily overcome, but only that it is "malleable." In saying that steel is "malleable," we do not mean that it can be easily bent, but only that, exposed to the proper amount of heat and pressure, it can be forged into, for instance, instruments and tools that enable human beings to transform their environment in a variety of ways. Soft constraints are ubiquitous in social life, and are also sometimes quite resistant to change. Soft constraints shape political life both within and across borders. They incline actors to choose certain means to the detriment of others in the pursuit of a wide range of goals, or to favor some public policies to the detriment of others in times of crisis. Soft constraints play a decisive role for the success—or for the failure, as the case may be—of issues of global concern such as climate goals (CGs) and pandemic goals (PGs). As of 2020, it seems safe to assume that there are no hard constraints on the feasibility of health policies to address the COVID pandemic effectively within the next three to four years, nor on the feasibility of CGs to avert dangerous climate change within the next thirty years. There emerges the question, then, as to which soft constraints stand in the way of both sets of goals (CG and PG), and how malleable they are. In order to address this question, we have to introduce a distinction that has become common in the climate debate, but that has been largely overlooked in the current debate on the COVID pandemic, namely a distinction between *adaption* and *mitigation* measures. There are, then, two sets of goals (PG and CG) and four subsets of goals, as shown in table 10.1.

The four subsets of goals can be described as follows:

SUBSET 1. adaptation goals within the set of pandemic goals (*P-A*)
SUBSET 2. mitigation goals within the set of pandemic goals (*P-M*)
SUBSET 3. adaptation goals within the set of climate goals (*C-A*)
SUBSET 4. mitigation goals within the set of climate goals (*C-M*).

Table 10.1 Pandemic goals and climate

	PG (Pandemic Goals)	CG (Climate Goals)
(A) adaptation	SUBSET 1 P-A soft-constraints	SUBSET 3 C-A soft-constraints
(M) mitigation	SUBSET 2 P-M soft-constraints	SUBSET 4 C-M soft-constraints

Some soft constraints may occur in more than one subset—or even in all of them. Consider, for instance, some of the legal, political, and economic constraints that governments had to overcome during the 2020 pandemic. These were soft-constraints on *P-A*, as the main goal of governments in 2020 was not to mitigate the occurrence of a new or future pandemic, but to adapt themselves to one that had already emerged. In order to contain the advance of COVID, governments had to impose temporary restrictions, for example, on air travel and freedom of movement. Governments also had to adjust their budget in order to protect people who were unable to work—whether as an employer or employee—during the health crisis. Now, in order to address CGs (subsets *C-A* and *C-M*) effectively, governments have to overcome similar legal, political, and economic constraints, though not temporarily as in subset *P-A*, but over a longer period of time (or perhaps even indefinitely). Global preparedness for future health emergencies involving the outbreak of new diseases (subset *P-M*), as we'll show in more detail later, also compels governments to overcome similar legal, political, and economic soft constraints over a long period of time. But some soft constraints, on the other hand, occur only in one specific subset. The subset *C-M* is a case in point here. The feasibility of *C-M* depends on trans- and intergenerational cooperation. This requires from the relevant actors a capacity to overcome a kind of soft constraint that we call a "time-related proximity constraint," which does not stand in the way of *P-A*, *P-M*, or *C-A*.

Although some soft constraints may occur in more than one subset, they usually have different degrees of malleability. One soft constraint may be more malleable, for instance, because the relevant actors believe that they have to overcome it only during a relatively short period of time, after which society will return to the status-quo-ante crisis and the soft constraint will recover its previous degree of malleability. We will use the word "malleable" to refer to a single property that soft constraints have: depending on the subset in which they occur, the relevant actors can overcome them with more or less difficulty, whether for a short period of time only or indefinitely. The first problem we intend to address in this chapter, then, consists in establishing the main soft constraints that occur in each subset. For the purpose of this chapter, we do not claim to have established all relevant soft constraints (there are dozens of them), but only those that are most salient for the feasibility of climate and pandemic goals. We will focus on two particularly robust soft constraints: "structural constraints" (or geopolitical constraints) and on what we call "proximity constraints." We distinguish two kinds of proximity constraints: space-related proximity constraints and time-related proximity constraints. The second problem we address in this chapter is why some soft constraints have proved more malleable in one subset (*P-A*) and far less malleable in the other subsets (*P-M*, *C-A*, and *C-M*). Being more malleable because the relevant actors do not expect to endure long-term sacrifices is not the only relevant factor at play here. We argue that structural constraints and time-related proximity constraints are very resistant to change (i.e., they are not very malleable).

II. MITIGATION AND ADAPTATION GOALS

Pandemics are not natural disasters like earthquakes, tsunamis, or volcanic eruptions. Pandemics, like climate change, have anthropogenic causes. It is well-known, for instance, that illegal wet markets can lead to virus spillover and, thus, spark the outbreak of a pandemic. Wildlife trade and encroachment on the habitat of wild species through deforestation (or through the fragmentation of forests) can also cause pathogens to spill over into human beings and, then, give rise to a pandemic. Improved affordability of air travel and increased movement of people across borders, too, contribute significantly to the rapid spread of new viruses (Dobson et al. 2020, 379; Tollefson 2020; Lindahl and Grace 2015). Writing in 2005, Michael Osterholm argued that the world was unprepared to deal with pandemics, in spite of clear evidence that pandemics were likely to become more frequent (Osterholm 2005). Since then, the scientific community and international monitoring bodies have called attention to the ever-increasing probability of new pandemics and to the importance of coordinated efforts to achieve *P-A* goals and *P-M* goals on a global scale (Pike et al. 2014; Clapper 2016, 13–14; WHO 2018; Coats 2019, 21). *P-M* aim at preventing the occurrence of new outbreaks. If an outbreak does occur, effective *P-M* can also attenuate the chances that it becomes a pandemic. Seen in this light, it is clear that the measures to address the COVID pandemic are not *P-M*, but *P-A*. They aim at reducing the impact of a disease that has already emerged and spread globally.

P-A are mostly local. States and municipalities have the authority to enforce them within their own borders. *P-A* do not require strong international cooperation (strong-I*N*C). *P-A* include, for instance, the enactment of emergency laws, the construction of field hospitals, the introduction of contact-tracing tools and large-scale testing, and the enforcement of quarantine and social distancing. These measures are supposed to be temporary. The quest for a vaccine, too, is an adaptation measure because its primary goal consists in adapting the human immune system to a new environment. The infrastructure and expertise deployed for the development of a vaccine now may eventually also indirectly strengthen *P-M* in the future. When the pandemic is eventually over, further vaccination of the population can be understood as a mitigation measure: it prevents a future pandemic. Both governments and citizens, then, have some strong incentive to pursue *P-A* by overcoming, for instance, legal and ethical soft constraints such as freedom of movement and concerns about violation of privacy, or to change temporarily their lifestyles by wearing facemasks and avoiding handshaking. But, paradoxically, when it comes to *P-M*, the same actors perceive the same soft constraints as less malleable, even considering that some studies, published prior to the COVID outbreak, show that *P-M* are far less costly than *P-A* (Pike et al. 2014; Madhav et al. 2018; see also UNEP 2020, 7). From a cost–benefit perspective, the sheer costs of *P-A*, when compared to the costs of *P-M*, provide good reasons to favor *P-M* over *P-A*. But from an ethical perspective, too, it is easy to recognize that *P-M* should take precedence over *P-A*, as the loss of lives owing to a major pandemic, such as that in 2020, cannot be adequately measured by means of cost–benefit analyses, nor can it be compensated for.

CG also requires mitigation and adaptation measures. *C-M* aim mainly at a drastic reduction of CO_2 emissions by 2030, and global climate neutrality by 2050 (IPCC 2018; Steininger, Meyer et al. 2020). *C-M* (like *P-M*) also requires strong-I*N*C. The benefits of *C-M* (similar to the benefits of *P-M*) are mostly global. However, the positive effects of *C-M* will take decades to materialize (Samset et al. 2020). The current generation, especially individuals who are in their forties or older, cannot expect to benefit significantly from *C-M* in their own lifetime. Effective *C-M*, thus, requires both strong-I*N*C and strong trans- and intergenerational cooperation (strong-I*G*C). *C-A*, on the other hand, like *P-A*, can be effective at a local level and within shorter time. *C-A* aims, for instance, at reshaping the infrastructure of cities to make them more robust against extreme weather, or building houses out of concrete rather than wood to make them less vulnerable to bushfires. *C-A* does not require strong-I*N*C. However, the longer-term effectiveness of *C-A* ultimately depends on the success of *C-M*. *C-A* without *C-M* is likely of no avail. Many countries with people living in coastal areas are likely to lack the economic means of implementing efficient adaptation measures, for example, if sea levels rise over 1 meter on average by 2070, or over 2 meters by the end of the twenty-first century (Bamber et al. 2019). In these countries, people will only have the option to retreat from the coastal lines.

During the 2020 pandemic, several soft constraints, which also stand in the way of CGs, have been successfully overcome, indirectly leading to a significant decrease in greenhouse gas emissions. There are reasons to believe, then, that the soft constraints on *P-A* are more malleable than the soft constraints on *P-M*, *C-A*, and *P-M*. But why? Our hypothesis is that the more a goal *G* requires strong-I*N*C and strong-I*G*C, the less malleable are the relevant soft constraints on *G*. This is systematized in table 10.2.

Table 10.2 International and intergenerational cooperation required for climate and pandemic goals

	P-A	*C-A*	*P-M*	*C-M*
strong-I*N*C	no	no	yes	yes
strong-I*G*C	no	no	no	yes

It is important to keep in mind here that although *C-A* requires neither strong-I*NC* nor strong-I*GC*, the long-term effectiveness of *C-A*, as we have emphasized earlier, does require *C-M*, which in turn requires both strong-I*NC* and strong-I*GC*. One might argue that effective *P-A* also requires strong-I*NC*, even if it does not require strong-I*GC*. After all, we cannot, for instance, expect each country to develop a vaccine for COVID by itself without the help of other countries. However, we can imagine at least three scenarios in which a country pursues *P-A* effectively without having to engage in strong-I*NC*. *Scenario 1*: A country (or a small coalition of countries) does indeed achieve a breakthrough in the development of an effective vaccine and only agrees to share the vaccine with other countries after its own population has been immunized, and provided it has stockpiled millions of doses for possible use in its own territory in the future. *Scenario 2*: A country (or a small coalition of countries) does not succeed in developing its own vaccine, but it is rich enough to negotiate with pharmaceutical companies' exclusive access to the vaccine before it can be distributed to poorer countries. *Scenarios 1* and *2* are instances of a practice known as "vaccine nationalism" (Kupferschmidt 2020; Brown and Susskind 2020; Callaway 2020). *Scenario 3*: A country does not manage to develop its own vaccine, nor does it have the financial resources to negotiate an exclusive supply agreement with pharmaceutical companies. This country, then, seals off its borders and enforces strict contact tracing and quarantine measures. In the process, thousands of citizens are killed due to lack of treatment and vaccine, but the remaining population eventually becomes immune or is never exposed to the virus because the borders are kept closed indefinitely as an attempt to preclude another health crisis in the future. In *Scenario 3*, a country would probably have to endure economic stagnation and sacrifice strong scientific or cultural partnerships with other countries. Now, of course *P-A* will be more effective at a global level if it is pursued through strong-I*GC*, but it does not necessarily require strong-I*GC*. On the other hand, no country can seal off its borders from the effects of dangerous climate change, or become entirely immune to its effects through the implementation of *C-A*. In order to avert dangerous climate change, both strong-I*NC* and strong-I*GC* are required.

What are, then, the most salient soft constraints on the feasibility of strong-I*NC* and strong-I*GC*? We argue that structural (or geopolitical) soft constraints and proximity soft constraints constitute the most robust constraints on the feasibility of both climate and pandemic goals. As we intend to show, proximity constraints can be even more robust than structural constraints.

III. STRUCTURAL SOFT CONSTRAINTS AND THE SYSTEM OF STATES

The structure of the system of states is such that states cannot rely on institutional bodies for the execution and enforcement of laws at a global level analogous to the bodies that typically exist in individual states at a national level. But it does not follow from this, of course, that the absence of a world government prevents states and international bodies from cooperating with one another in a wide range of activities. States and international bodies engage in diplomacy, trade, international regimes, academic partnerships, humanitarian aid, and several other forms of cooperative enterprises for the sake of mutual benefits. Consider, for instance, the global effort to eradicate smallpox in the 1970s. It is a testimony to humankind's capacity to cooperate across state borders over a long period of time, about ten years in this case. Similarly, polio eradication has been often hailed as "a triumph of international cooperation and of preventive medicine" (Henderson 1980, 5), or as "perhaps the greatest achievement of international cooperation in human history" (Barrett 2016, 14519). Several authors in the field of international relations argue that international institutions, such as the World Health Organization, have played a crucial role for the success of these sorts of achievements. For these authors, sometimes referred to as supporters of liberal institutionalism, states are not the only relevant actors in the international arena and the structural soft constraints resulting from the absence of a world government do not preclude international cooperation (see e.g., Keohane 1984; Weiss 2013).

Supporters of liberal institutionalism can explain the success of projects such as the eradication of polio and the endurance of institutions such as diplomacy and humanitarian aid. Yet, international institutions have not been able to prevent the outbreak of the new coronavirus from becoming a major pandemic, nor have international institutions hitherto prevented the increase of greenhouse gas emissions in the atmosphere over the last decades. Why?

Supporters of realism in international relations will tend to hypothesize that structural constraints are less malleable than supporters of liberal institutionalism tend to argue. According to realism, states are the main, if not the only, relevant actors in the context of international relations. In some critical areas of interest, a state has to rely on self-help, rather than on international institutions, in order to promote its internal security. As Kenneth Waltz, a well-known supporter of realism, famously put the problem: "A national system is not one of self-help. The international system is" (Waltz 1979, 104). Each state, understood as a sovereign political body, is ultimately responsible for its own security. Now, one realist explanation for lack of strong-INC to pursue *P-M* and (especially) *C-M* is that, in the absence of a world government, each state perceives its individual economic performance in the short run as more critical to its security than the prospect of averting pandemics and climate change in the long run. One reason for this is the assumption that, from a realist perspective, there is too much uncertainty underlying the economic costs of unmitigated climate change, whereas the costs of effective *C-M* are knowingly very high—high enough to halt economic growth and disturb internal stability (Nordgren 2016, 1045; Symons 2018, 5; Mearsheimer 2001, 371). The problem is compounded by the game-theoretical insight that today for most, if not all states, considered individually, it is the case that each is better-off if it reaps the benefits of strong-I*N*C, pursued by the other states, without itself sharing in the costs of *C-M* (Nordgren 2016, 1047).

Of course, proponents of liberal institutionalism will not deny that state behavior often reflects the logic of egoistic self-preservation and the maximization of state welfare, however understood. Both schools of international relations—realism and liberal institutionalism—agree that structural constraints may explain the behavior of states within the international arena. But proponents of liberal institutionalism argue that because states are not the only relevant actors in the international arena, structural constraints can often be overcome by means of diplomacy, international institutions, and non-governmental bodies. According to supporters of realism, on the other hand, structural constraints have greater explanatory power than the supporters of liberal institutionalism are willing to admit. Structural constraints, the realists argue, explain how states have behaved in the past; why states prioritize, for instance, economic growth over CGs in the present; or how states, in the face of uncertainties, are expected to act in future. And some supporters of realism take even a step further and argue that realism not only explains the behavior of states but also prescribes how states *should* act in their struggle for survival in a world devoid of central government. John Mearsheimer refers to this branch of realism as "offensive realism":

> It explains how great powers have behaved in the past and how they are likely to behave in the future. But it is also a prescriptive theory. States should behave according to the dictates of offensive realism, because it outlines the best way to survive in a dangerous world. (Mearsheimer 2001, 11)

Structural constraints deliver a plausible account of the behavior of states during the 2020 pandemic. The reason most governments were quick to implement drastic measures against the spread of the new coronavirus (*P-A*), while remaining slow to pursue *P-M*, is that they were primarily concerned with the consequences of the pandemic within their own borders. In the face of a major crisis, states must rely on self-help, rather than on international institutions, to promote internal security. When the chips are down, the state is the only actor that has the legitimate power, for instance, to ground long-distance flights, to enforce lockdown and quarantine on its own population, or to impose temporary restrictions on basic rights such as freedom of movement and assembly. International institutions can provide helpful advice, but they have neither the power nor the legitimacy to implement and coordinate effective *P-A* at a national level. Let us suppose, then, for the sake of argument, that this is a sound description of the reasons that underlie the behavior of states in the midst of a major health crisis. Should we conclude from this that states *should* also act within the bounds of the structural constraints we've outlined in the face of global risks in future?

From the fact that states have pursued a certain pattern of behavior in the past, we must not infer that states *should* stick to that pattern of behavior in the future. Supporters of realism favor internal security over strong-I*N*C because they consider the struggle for power the only way to survive in a world devoid of central government. As Mearsheimer puts the problem in his influential defense of realism: "[. . .] my theory sees great powers as concerned mainly with figuring out how to survive in a world where there is no agency to protect them from each other; they quickly realize that power is the key to their survival" (Mearsheimer 2001, 20). This approach might sound reasonable in scenarios in which the main threat to the security of a state is the existence of other states. But is it still attractive in the face of threats such as the increased frequency of pandemics and dangerous climate change? Mearsheimer's defense of offensive realism involves the assumption that we live in a "dangerous world," but the kind of danger posed by climate change and future pandemics cannot be addressed by means of offensive realism, for the nature of these threats is such that they cannot be overcome without strong-I*N*C. In spite of the vocabulary of war that heads of state have used at the outset of the global health crisis in 2020, it is clear that climate change and pandemics—to mention only two obvious global threats—represent a different kind of enemy. Even influential realists such as Hans Morgenthau seem to have realized, early in the 1960s, that the system of states is ill-suited to address threats that pose global risks, and which can only be addressed by means of strong-I*N*C involving virtually all states, especially the most powerful ones. In the wake of the first tests with thermonuclear bombs, he wrote the following:

> Modern technology has rendered the nation state obsolete as a principle of political organization; for the nation state is no longer able to perform what is the elementary function of any political organization: to protect the lives of its members and their way of life. [. . .] Under the technological conditions of the pre-atomic age, the stronger nation states could, as it were, erect walls behind which their citizens could live in safety while the weaker states were protected by the operation of the balance of power, which added the resources of the strong to those of the weak.

The modern technologies of transportation, communications, and warfare, and the resultant feasibility of all-out atomic war, have completely destroyed this protective function of the nation state. No nation state is capable of protecting its citizens and their way of life against an all-out atomic attack. Its safety rests solely in preventing such an attack from taking place. (Morgenthau 1966, 9)

The scenario Morgenthau had in mind was, in a certain sense, even more optimistic than our predicament in the face of climate change, for the states are not under a time constraint to make a nuclear war impossible by effectively banning nuclear weapons. Under the doctrine of mutual deterrence, nuclear powers can, at least in principle, indefinitely maintain their nuclear arsenals and remain committed to using them only in retaliation for a nuclear attack. The relevant time constraint here, therefore, is not about how to avert a nuclear war, but how to react in the event of a nuclear attack. If a nuclear power has some reason to believe it is on the verge of being hit by nuclear weapons, it has to decide within minutes, or perhaps even seconds, whether or not to retaliate. A major nuclear conflict, thus, can arise on the grounds of pressing time constraints and poor evidence for retaliation (Cirincione 2008, 383–384). The time constraint to meet current CGs, on the other hand, is also quite pressing. But the reasons to act quickly in order to avert dangerous climate change are not based on poor evidence. A growing body of evidence shows that delaying the implementation of effective *C-M* will only exacerbate the costs, including human costs, resulting from dangerous climate change (Working Group III 2014, 12). Thus, we cannot expect to address the kind of danger that climate change poses by simply following the prescriptions of offensive realism.

Structural constraints, though, especially as they are typically perceived by supporters of realism, remain as one of the most robust obstacles in the way of effective climate polices. On the other hand, it is important to keep in mind that structural constraints, as supporters of liberal institutionalism have emphasized over the last decades, are not hard constraints. They are subject to change. And as we intend to suggest later in this chapter, there is, indeed, some reason to believe that structural constraints may become more malleable in a post-pandemic world.

IV. PROXIMITY CONSTRAINTS AND THE CLAIMS OF FUTURE PEOPLE

There are other kinds of relationships that give rise to robust soft constraints on the feasibility of both climate and pandemic goals. The unwavering support some people give to their own state, even when the actions of one's own state are deleterious to the interests of citizens of other states, can easily stand in the way of strong-I*NC*. Sometimes, on the other hand, people tend to favor the interest of people of their own generation—whether or not they happen to be their co-nationals—neglecting the interests of people of the next generations. Or sometimes some people favor, rather, the interests of other people, not because they are their co-nationals or their contemporaries, but because they share the same religion or the same worldview, or simply because they are their friends or family members. These different kinds of relation—and we have mentioned just a few examples here—give rise to ties of loyalties that, in many respects, may be laudable, as they promote mutual support and foster a sense of community. But these ties do also often stand in the way of strong-I*NC* and strong-I*GC*. We call these relations "proximity constraints."

Proximity constraints motivate actors to be partial toward the well-being of people who are *close* to them. Proximity constraints affect the behavior of actors over space and time. The spatial dimension is relatively clear: we tend to favor the interests of people who are spatially close to us. But it is important to spell out the spatial dimension of proximity constraints in some more detail in order to avoid confusion. One may feel close to someone or to some people, while not being spatially close to them in a literal sense. A Muslim émigré, for instance, may experience bonds of loyalty to fellow Muslims even while not actually living in the same territory they do. Friends and family members may believe they have special duties toward each other even while living thousands of miles apart. Proximity constraints usually originate on the grounds of physical proximity, but the sense of obligation and allegiance they create may remain in place long after physical separation, past emigration, or a historical diaspora. We speak of space-related proximity constraints here to refer to perceived proximity, or a sense of closeness, rather than actual or physical proximity.

One obvious instance of space-related proximity constraints is nationalism. Henry Shue refers to this particular kind of soft constraint as the "compatriots take priority" principle (Shue 1980, 131–132). More recently, the same attitude has been referred to as the "my country first" approach (Brown and Susskind 2020, S68; Contractor 2017). During a major health crisis, a state can overcome some space-related proximity constraints at a domestic level by enforcing measures that, for example, prevent citizens, keen on protecting their own families and friends first, from stockpiling essential goods. But preventing states from benefiting their own people, even to the detriment of people in other states, is much harder. During the 2020 pandemic, states did not have to overcome nationalism because this particular kind of space-related proximity constraint was, rather, instrumental to the feasibility of *P-A*. Nationalism enabled some states, for example, to justify a ban on the export of food and medical equipment in order to promote internal security during the pandemic. The American government, for instance, was accused of illegally intercepting European supplies of personal protective equipment, used by frontline healthcare workers, in order to benefit its own population. And long before a vaccine for COVID could be considered effective and safe, some analysts had already been warning that vaccine nationalism might delay the immunization of the world population, as some states, or group of states, might try to stockpile doses or control the supply of raw ingredients to benefit their own people first, even at the cost of populations in other countries (Bollyky and Bown 2020; Kupferschmidt 2020; Brown and Susskind 2020). Rather than being an obstacle, proximity constraints such as nationalism can actually promote the feasibility of *P-A* (and possibly of *C-A* as well). But in the long run, adaptation measures to pandemics and to the climate crisis are of no avail, and much costlier, if they are not accompanied, respectively, by mitigation measures to pandemics and to the climate crisis. But while mitigation measures to pandemics require strong-I*N*C, mitigating the climate crisis requires strong-I*N*C and strong-I*G*C. This leads us to the examination of the *temporal* dimension of proximity constraints.

The temporal dimension may perhaps be less apparent than the spatial dimension, but it is even more pressing for the assessment of the feasibility of CGs: we tend to favor the interests of people who are close to us in time over those of more remote future people and sometimes even our own immediate interests over our more important future interests. The non-reciprocal nature of the relationship between present and future people is such that while currently living people can affect the conditions of life of future people, future people can do nothing to undermine the welfare of currently living people. The inequality of power in the relationship between currently living people and future generations leads to what Stephen Gardiner aptly calls "the tyranny of the contemporary" (Gardiner 2011, 143–184).

One might suggest, however, that in the case of CG the tyranny is only apparent, and that to speak of tyranny is quite misleading, for the very existence of future people depends on the actions that previous generations have pursued in the past. Thomas Schwartz and Derek Parfit are among the first philosophers who drew attention to this kind of paradox (Parfit 1976; Parfit 1984, 351–390; Schwartz 1978). It is a paradox because future people might perhaps want to blame past generations (the current generation) for having engaged in activities that are deleterious to them in the future, but the matter of fact is that without the actions of previous generations those people in the future would not exist as the particular individuals they are. Different courses of action in the past would have brought about the birth of different persons in the future, for their parents, or grandparents, or great grand-parents would probably never have met without those particular courses of action, which might have been avoided. Schwartz called this paradox the "fallacy of beneficiary-conflation" (Schwartz 1978, 7), but it is now better known as the "non-identity-problem" (Parfit 1984, 359). The philosophical literature on the non-identity-problem has become quite extensive, and the threshold notion of harm offers one of the successful responses to the problem (Shiffrin 1999; Meyer 2003; Meyer 2020, sect. 3.1). Also, on the basis of the interest-theory of rights (Raz 1994, 45–51; Kramer 1998, 60–101) future people can have rights vis-à-vis currently living people. Being able to exercise one's rights—to demand or waive the enforcement of a right—is neither sufficient nor necessary for someone to be the bearer of the right (Meyer 2020, sect. 1). We can, for instance, attribute rights to people who are not in a position to claim the fulfillment of their interests and needs such as, for instance, infants, mentally disabled persons, and even non-human animals. We attribute rights to them on the assumption that they are subject to suffering, or that they have interests, basic needs, or capacities that prompt us to attribute similar rights to actually living persons. There is wide agreement now, among both normative theorists and policymakers, that skepticism about the moral force of the claims of future people vis-à-vis currently living people is not tenable. The 2014 Assessment Report of the United Nations Intergovernmental Panel on Climate Change (IPCC) also acknowledges that these theoretical issues have already been adequately addressed (Working Group III 2014, 215–218).

At this point, one might perhaps be willing to agree that future people do have valid moral claims vis-à-vis currently living people. But one might also suggest that due to the distance in time, our concern for future people counts less than our concern for currently living people (Nordgren 2016, 1047). The force of their moral claims, one might add, becomes weaker and weaker the farther in time future generations are from us. But is this suggestion fair? Does the force of moral claims fade in the course of time? Being indifferent to the interests of future people, as Frank Ramsey put the problem nearly a hundred years ago, is "ethically indefensible and arises merely from the weakness of the imagination" (Ramsey 1928, 543). The same point had already been made by Henry Sidgwick, who famously argued that:

> The time at which a man exists cannot affect the value of his happiness from a universal point of view; and that the interests of posterity must concern a Utilitarian as much as those of his contemporaries, except in so far as the effect of his actions on posterity—and even the existence of human beings to be affected—must necessarily be more uncertain. (Sidgwick 1907, 414)

And one does not have to endorse utilitarianism, as Sidgwick did, in order to recognize that moral concern for the well-being of future people does not diminish due to distance in time. As Rawls puts the point in *A Theory of Justice*: "The mere difference of location in time, of something's being earlier or later, is not a rational ground for having more or less regard for it" (Rawls 1971, 293; see also Parfit 1984, 367, 480–486). To assume that the interests and needs of future people weigh less due to distance in time is as arbitrary as to assume that the interests of the poor, in other parts of the world, or within one's own state, should count less simply because they are geographically distant from us. The reasons currently living people have to be concerned about the well-being of their contemporaries are no less compelling than the reason they have to be concerned about the well-being of people who are going to live in the future.

Some theorists such as John Broome, Eric Posner, and Cass Sunstein have argued, on the other hand, that effective measures to prevent dangerous climate change do not necessarily come with net-burdens for the current generation (Broome 2018; Posner and Sunstein 2008). They suggest that *C-M* can be implemented in a universally pareto-optimal way so that the current generation could shift the costs of *C-M* onto future generations. Everyone is better-off, they argue. However, as some theorists have pointed out, this proposal is clearly indefensible on normative (Brennan and Sayre-McCord 2016, 440–441), theoretical (Kelleher 2015, 75–77), and practical grounds (Caney 2014, 133–134). The feasibility of CGs, therefore, depends on our capacity to address both the spatial and the temporal dimensions of soft proximity constraints. For now, though, these constraints remain very robust.

V. GLOBAL RISKS IN A POST-PANDEMIC WORLD

Some soft constraints may be very robust, but their force may change in the course of time. Soft constraints, in this sense, are "time-sensitive." As Gilabert and Lawford-Smith put the problem: "it is possible to transform or dissolve them so that they are no longer constraints at some future time" (Gilabert and Lawford-Smith 2012, 814). Given the unprecedented human and economic costs resulting from the 2020 pandemic, and the probability of new outbreaks in the future, couldn't we expect structural and proximity constraints, some of the main hurdles on the feasibility of both climate and pandemic goals, to "dissolve," or at least to become more malleable in a post-pandemic world? We would like to submit two hypotheses regarding the future of global risks. The first hypothesis is that in spite of robust structural and spatial proximity constraints, *P-M* will be seen as a matter of priority in the international arena over the next years. Policymakers and citizens around the world will become increasingly inclined to overcome the structural and space-related proximity constraints that stand in the way of effective *P-M*. The history of international law suggests that some important new legal frameworks have been created as a direct response to catastrophic events. Nuclear accidents such as those in Chernobyl (1986), and in Fukushima (2011), for instance, prompted the International Atomic Energy Agency (IAEA) and individual states to revise domestic legislation and international treaties for the safe use of nuclear energy (IAEA 2006, 76; Stephens 2016, 167; Gioia 2012, 100; Tromans 2010, 253). The second hypothesis is that increased international cooperation to prevent other pandemics may lead to increased international responsiveness to the threats posed by dangerous climate change. We would like to spell out both hypotheses in some more detail, and address some objections that can be raised against them.

People who live in a hypothetical country A, quite distant from country B, might be willing to pay the citizens of B to adopt measures that minimize the chances that a virus spillover in country B evolves into a pandemic. The payment can be mediated, and to a certain extent does already exist, by means of institutions such as the World Health Organization. The reasons the citizens of A have to act for the benefit of B are clear: the citizens of A may not care much about what happens in B, for they do not feel close to them, but the citizens of A care a lot about what happens in A. Even a die-hard realist, or a right-wing nationalist, then, might support commitment to strong-I*N*C and recognize that state borders, while important to keep human enemies at bay, are of little help in the attempt to prevent future pandemics. State borders did not prevent the most powerful state, under the leadership of a president clearly committed to the precepts of realism and an ardent supporter of the "my country first" approach, to become the state with greatest number of deaths resulting from COVID—more deaths, indeed, than the combined number of causalities it inflicted in Hiroshima and Nagasaki in 1945. The 2020 pandemic alone caused more deaths in the United States than the combined number of casualties of American soldiers in the Gulf War (1990–1991), in the War in Afghanistan (2001–present), and in the Iraq War (2003–2011). It takes only one infected co-national to bring a virus home and begin the spread of a new disease throughout an entire country. If health authorities are quick to detect the problem, closing the borders may help to slow down the advance of new infections, but it can hardly prevent the virus from entering a country in the first place. The National Academy of Medicine, an American non-governmental organization, summarized the problem in a report published in 2016: "Pandemics know no borders, so international cooperation is essential. Global health security is a global public good requiring collective action" (National Academy of Medicine 2016, 4).

The vivid experience of the 2020 pandemic may lead citizens and policy-makers to put a different weight on the force of space-related proximity constraints and structural constraints, especially if the general public also becomes more vividly aware that scientists, research groups, and international bodies have been warning about the imminence of a major pandemic for years. In 2016, for instance, the United States Intelligence Community, which provides advice to the American Senate in matters related to global security, produced a report that concluded that the "international community remains ill prepared to collectively coordinate and respond to disease threats," including coronaviruses (Clapper 2016, 13–14). Early in 2019, just months before the first case of COVID was reported in China, the United States Intelligence Community published a new report on global threats and suggested, again, that the international community remained ill-prepared to address the outbreak of new pandemics. This time the group also highlighted that climate change can aggravate the occurrence of new outbreaks:

> We assess that the United States and the world will remain vulnerable to the next flu pandemic or large-scale outbreak of a contagious disease that could lead to massive rates of death and disability, severely affect the world economy, strain international resources, and increase calls on the United States for support. Although the international community has made tenuous improvements to global health security, these gains may be inadequate to address the challenge of what we anticipate will be more frequent outbreaks of infectious diseases because of rapid unplanned urbanization, prolonged humanitarian crises, human incursion into previously unsettled land, expansion of international travel and trade, and regional climate change. (Coats 2019, 21)

Also in 2019, the Global Preparedness Monitoring Board, an independent monitoring and advocacy body, issued a similar warning: "there is a very real threat of a rapidly moving, highly lethal pandemic of a respiratory pathogen killing 50 to 80 million people and wiping out nearly 5% of the world's economy. A global pandemic on that scale would be catastrophic, creating widespread havoc, instability and insecurity" (Global Preparedness Monitoring Board 2019, 6). As Morten Broberg pointed out, there are plenty of reports and academic studies, published over the years leading up to the 2020 global health crisis, that called attention to the imminence of pandemics and to the world's lack of preparedness to deal with global health issues (Broberg 2020, 204; see also Maxmen and Tollefson 2020). Let us suppose, then, as we have suggested, that public awareness of past warnings may make civil society and policymakers more vividly aware that the 2020 pandemic was not a one-off event, and that strong-I*NC* is necessary in order to implement both *P-A* and *P-M*. What is the implication, then, for our discussion on the feasibility of CGs?

There is now a solid body of evidence on the catastrophic consequences of unmitigated climate change, and on the adaptation and mitigation measures that should be undertaken in order to address the problem effectively, just as there was, too, a reliable body of evidence that strongly suggested, as early as 2005, that a pandemic like that of 2020 was going to strike sooner or later. As the effects of the 2020 pandemic recede over the next years (or perhaps decades), younger people around the world, who were perhaps not even entitled to vote as the global health crisis broke out, may come to believe that re-establishing the status-quo-ante is not their only or best option. A recent survey in 14 countries shows, for instance, that the younger generation is increasingly aware of the perils associated with unmitigated climate change; they consider climate change even more threatening than possible future pandemics (Poushter and Huang 2020). The consequences of business-as-usual attitudes in our relationship to the environment may turn out to be perceived, then, as a real existential threat. A new generation of legislators and magistrates may also be appointed to key offices in the decision-making processes (Schreuers and Papadakis 2020, 16–24). For these reasons, the next generation may become willing to sacrifice at least some aspects of their lifestyles in order to address global risks in the future, and their political representatives, too, might turn out to be seriously committed to pursuing the relevant policies.

This is, of course, a bold hypothesis, one that some people would rather dismiss as too naively optimistic to be true. One might want to remind us that in one important respect *P-M* are quite different from *C-M*, as the former, unlike the latter, do not depend on strong-I*GC*. This means that each generation may perhaps be willing to support *P-M* because they themselves expect to benefit from these measures, especially if they have witnessed events such as the 2020 pandemic. But when it comes to *C-M*, each generation may prefer, instead, to buck-pass the costs of mitigation and adaptation measures to future generations. Thus, even if the next generation becomes more willing to overcome the structural constraints that stand in the way of strong-I*NC*, which is necessary to achieve *P-M*, it would still be necessary to overcome the time-related proximity constraints that stand in the way of strong-I*GC*, which is necessary to avert dangerous climate change.

Thus, even a supporter of political realism might be willing to admit that, as matter of self-interest, states should perceive pandemics as a real threat to their internal security and that, under these circumstances only, states might become willing to endorse, or perhaps go as far as to enforce, cross-border legal frameworks and international regimes to avert new disease outbreaks that might evolve into a pandemic. There is, indeed, some reason to believe that the 2020 pandemic may already have weakened the force of the structural constraints on the feasibility of *P-M*. In 2020, three unabashed supporters of the "my country first" approach have tested positive for COVID: British prime minister Boris Johnson, Brazilian president Jair Bolsonaro, and American president Donald Trump. Some members of their respective staff have also been infected. In the United States, as reported by the magazine *The*

New Yorker, "The entire Joint Chiefs of Staff—the highest-ranking officers from the Army, Navy, Air Force, Coast Guard, and other services—went into quarantine for two weeks" (Wright 2020). According to the magazine, some leading American strategists feared that some competing states, seeing this as a vulnerability, might take advantage and attempt to threaten American interests abroad. The magazine suggests the following: "From multiple angles, including the fact that President Trump is also the Commander-in-Chief, the coronavirus is now a genuine national-security threat for the United States" (Wright 2020). A supporter of realism might recognize, then, that as a matter of "enlightened self-interest" (Attfield 2014, 183; Symons 2018, 12), each individual state has good reasons to pursue not only *P-A* at a domestic level but also *P-M* at an international level. But a supporter of realism might also suggest that the feasibility of *P-M* requires mutual surveillance and a great deal of pressure over states that threaten to flout the rules. However, the nature of the relationship between current and future generations is such that future people can neither oversee the actions of currently living people nor exert pressure on them. For this reason, a supporter of realism might still see no reason to engage in strong-I*N*C for the purpose of *C-M*. Nonetheless, we would like to conclude by adding two points that support our optimistic hypothesis, which, as suggested earlier, reflects the following considerations: During the COVID-19 pandemic many, if not most people perceived the economically costly measures that significantly restricted their freedoms as justified. They considered these measures as necessary to contain the COVID-19 pandemic successfully. The experience of this collective burdensome and successful effort may increase their willingness to accept similar burdens and costs in responding to other global risks, including climate change. Accordingly, they may want to support the relevant political actors in overcoming the constraints that need to be overcome to avoid a climate catastrophe.

We conclude with two points that support the optimistic hypothesis. The first point is that climate change has geopolitical implications that have a great potential to foster international conflicts. This might motivate the relevant actors to overcome the structural constraints on the feasibility of *C-M* in order to avoid disruption to the current balance of power. In 2014, the United States Department of Defense published a document in which it refers to the effects of climate change as "threat multipliers":

> Climate change may exacerbate water scarcity and lead to sharp increases in food costs. The pressures caused by climate change will influence resource competition while placing additional burdens on economies, societies, and governance institutions around the world. These effects are threat multipliers that will aggravate stressors abroad such as poverty, environmental degradation, political

instability, and social tensions—conditions that can enable terrorist activity and other forms of violence. (Hagel 2014, 8)

These are, of course, geopolitical reasons to support long-term cooperation for the sake of *C-M*. But the attempt to foretell international conflicts is certainly not the only reason, not even the most pressing reason, to support *C-M*. Geopolitical reasons will not help us overcome the time-related proximity constraints, which according to our analysis remain as one of the most robust constraints on the feasibility of *C-M*. This leads us to the second point.

The second point is the following: we are not arguing that, on balance, we have more reasons to believe that humanity will implement effective *C-M* in time to prevent catastrophic climate change than we have to believe that, in the end, CGs are not feasible. We are putting forward reasons to foster what Beardsworth calls "the mindset of a fierce optimism" (Beardsworth 2020, 386). In hypothesizing that stronger international responsiveness to future outbreaks may spill over to increased responsiveness to the threats posed by dangerous climate change, and to greater sensibility toward the claims of future people, we do not claim that this is the most plausible prognosis of future developments in view of the best evidence available, rather we are suggesting that the current generation can avoid past mistakes and respond to the climate crisis responsibly. But, one might ask, is it a valid move in the course of a philosophical discussion on the feasibility of CGs? Yes, at least as long as we make it clear that we are not simply trying to predict the outlines of a rosier future, but first and foremost trying to put forward reasons to preclude the emergence of a catastrophic one. One objection that has been often posed against political realism is that it starts as a descriptive theory on the behavior of states, but then it rapidly evolves into a normative theory on how the states should behave within a system devoid of central government. Political leaders and policymakers, then, assume the theory is true and try to adjust their behavior to what the theory purports to predict. Realism, as many authors have realized, gains support from its self-fulfilling prophecies (Wendt 1992, 410). Now, one might finally object that our optimistic hypothesis shares with political realism the hope of being a self-fulfilling prophecy. We do not deny this.

In the preface to *Essays in Persuasion*, published in 1931, John Maynard Keynes famously affirmed: "For if we consistently act on the optimistic hypothesis, this hypothesis will tend to be realised; whilst by acting on the pessimistic hypothesis we can keep ourselves for ever in the pit of want" (Keynes 1931). If this is so, we are willing to accept the final objection to our optimistic hypothesis as a hopeful contribution to the materialization of a less turbulent, more hospitable future, for generations to come.[1]

NOTE

1. We would like to thank our colleagues at Graz University, section Moral and Political Philosophy, for their most valuable comments and criticisms on various drafts of the text, and in particular, Deborah Biging, Elisa Moser, Norbert Paulo, Daniel Petz, Thomas Pölzler, and Santiago Truccone. Our sincere thanks also to Kian Mintz-Woo (University College Cork) for sharing his invaluable suggestions and criticisms. We had the opportunity to present the penultimate draft of the chapter at the 4th UK-Latin America Political Philosophy Network (Uklappn) on November 14, 2020. Felicitas Holzer (University of Zürich) served as our commentator and raised a good number of central and critical questions. We would also like to thank the workshop participants for the lively and constructive discussion, and, in particular, Francisco Garciá Gibson (University of Buenos Aires) for his written comments and criticisms. Last, but not least we would like to thank the editors of the volume, Corey Katz and Sarah Kenehan. They very much encouraged us in working out the chapter and both provided detailed comments and criticisms to various drafts of the text that helped us enormously.

REFERENCES

Attfield, Robin. 2014. *Environmental Ethics: An Overview for the Twenty-First Century*. Cambridge: Polity Press.

Bamber, Jonathan L., Michael Oppenheimer, Robert E. Kopp, Willy P. Aspinall, and Roger M. Cooke. 2019. "Ice Sheet Contributions to Future Sea-Level Rise from Structured Expert Judgment." *Proceedings of the National Academy of Sciences USA* 116(23): 11195–11200.

Barrett, Scott. 2016. "Coordination vs. Voluntarism and Enforcement in Sustaining International Environmental Cooperation." *PNAS (Proceedings of the National Academy of Sciences)* 113(51): 14515–14522.

Beardsworth, Richard. 2020. "Climate Science, the Politics of Climate Change and Futures of IR." *International Relations* 34(3): 374–390.

Bollyky, Thomas J., and Chad P. Bown. 2020. "The Tragedy of Vaccine Nationalism. Only Cooperation Can End the Pandemic." *Foreign Affairs*, July 27. https://www.foreignaffairs.com/articles/united-states/2020-07-27/vaccine-nationalism-pandemic.

Brennan, Geoffrey, and Geoffrey Sayre-McCord. 2016. "Do Normative Facts Matter . . . to What is Feasible?" *Social Philosophy and Policy* 33(1/2): 434–456.

Broberg, Morten. 2020. "A Critical Appraisal of the World Health Organization's International Health Regulations (2005) in Times of Pandemic: It is Time for Revision." *European Journal of Risk Regulation* 11(2): 202–209.

Broome, John. 2018. "Efficiency and Future Generations." *Economics and Philosophy* 34(2): 221–241.

Brown, Gordon, and Daniel Susskind. 2020. "International Cooperation During the COVID-19 Pandemic." *Oxford Review of Economic Policy* 36(Supplement 1): S64–S76.

Callaway, Ewen. 2020. "The Unequal Scramble for Coronavirus Vaccines – By the Numbers. Wealthy Countries Have Already Pre-Ordered More Than Two Billion Doses." *Nature* 584(7822): 506–507.
Caney, Simon. 2014. "Two Kinds of Climate Justice: Avoiding Harm and Sharing Burdens." *Journal of Political Philosophy* 22(2): 125–149.
Cirincione, Joseph. 2008. "The Continuing Threat of Nuclear War." In *Global Catastrophic Risks*, edited by Nick Bostrom and Milan M. Cirkovic, 381–401. Oxford: Oxford University Press.
Clapper, James. 2016. *Statement for the Record Worldwide Threat Assessment of the US Intelligence Community*, 9 February. https://www.armed-services.senate.gov/imo/media/doc/Clapper_02-09-16.pdf.
Coats, Daniel R., ed. 2019. *Statement for the Record Worldwide Threat Assessment of the US Intelligence Community*, January 29. https://www.dni.gov/files/ODNI/documents/2019-ATA-SFR---SSCI.pdf.
Contractor, Farok J. 2017. "Global Leadership in an Era of Growing Nationalism, Protectionism, and Anti-Globalization." *Rutgers Business Review* 2(2): 163–185.
Dobson, Andrew P., Stuart L. Pimm, Lee Hannah, et al. 2020. "Ecology and Economics for Pandemic Prevention: Investments to Prevent Tropical Deforestation and to Limit Wildlife Trade Will Protect Against Future Zoonosis Outbreaks." *Science* 369(6502): 379–381.
El Zowalaty, Mohamed E., Sean G. Young, and Josef D. Järhult. 2020. "Environmental Impact of the COVID-19 Pandemic—A Lesson for the Future." *Infection Ecology & Epidemiology* 10(1): 1768023.
Gardiner, Stephen. 2011. *A Perfect Moral Storm*. New York: Oxford University Press.
Gilabert, Pablo, and Holly Lawford-Smith. 2012. "Political Feasibility: A Conceptual Exploration." *Political Studies* 60: 809–825.
Gioia, Andrea. 2012. "Nuclear Accidents and International Law." In *International Disaster Response Law*, edited by A. de Guttry, M. Gestri, and Venturini, 85–102. The Hague, The Netherlands: Springer.
Global Preparedness Monitoring Board. 2019. *A World at Risk: Annual Report on Global Preparedness for Health Emergencies*. Geneva: World Health Organization. https://apps.who.int/gpmb/assets/annual_report/GPMB_annualreport_2019.pdf.
Hagel, Chuck (Secretary of Defense). 2014. *The Quadrennial Defense Review (QDR)*. United States: Department of Defense. https://history.defense.gov/Historical-Sources/Quadrennial-Defense-Review/.
Henderson, D. A. 1980. "A Victory for All Mankind." *The World Health (The Magazine of the World Health Organization)*, May: 3–5.
IAEA (International Atomic Energy Agency). 2006. *International Nuclear Law in the Post-Chernobyl Period (A Joint Report by the OECD Nuclear Energy Agency and the International Atomic Energy Agency)*. Paris: OECD.
IPCC. 2018. *Global Warming of 1.5°C: An IPCC Special Report on the Impacts of Global Warming of 1.5°C Above Pre-Industrial Levels and Related Global Greenhouse Gas Emission Pathways, in the Context of Strengthening the Global Response to the Threat of Climate Change, Sustainable Development, and Efforts*

to Eradicate Poverty, edited by V. Masson-Delmotte, P. Zhai, H.-O. Pörtner, D. Roberts, et al. In Press. https://www.ipcc.ch/sr15/.
Jewell, Jessica, and Aleh Cherp. 2019. "On the Political Feasibility of Climate Change Mitigation Pathways: Is It Too Late to Keep Warming Below 1.5°C?" WIREs Climate Change 11: e621.
Kelleher, J. Paul. 2015. "Is There a Sacrifice-Free Solution to Climate Change?" Ethics, Policy & Environment 18(1): 68–78.
Keohane, Robert. 1984. After Hegemony: Cooperation and Discord in the World Political Eeconomy. Princeton: Princeton University Press.
Keynes, John Maynard. 1931. "Essays in Persuasion." https://www.economicsnetwork.ac.uk/archive/keynes_persuasion/.
Kramer, Matthew H. 1998. "Rights Without Trimmings." In A Debate over Rights, edited by Matthew H. Kramer, Nigel E. Simmonds, and Hillel Steiner, 7–111. Oxford: Clarendon Press.
Kupferschmidt, Kai. 2020. "'Vaccine nationalism´ Threatens Global Plan to Distribute COVID-19 Shots Fairly." Science, 28 July.
Lawford-Smith, Holly. 2013. "Understanding Political Feasibility." The Journal of Political Philosophy 21(3): 243–259.
Le Quéré, C., R. B. Jackson, M. W. Jones, et al. 2020. "Temporary Reduction in Daily Global CO_2 Emissions During the COVID-19 Forced Confinement." Natural Climate Change 10: 647–653.
Lindahl, Johanna F., and Delia Grace. 2015. "The Consequences of Human Actions on Risks for Infectious Diseases: A Review." Infection Ecology and Epidemiology 5.
Madhav, Nita, Ben Oppenheim, Mark Gallivan, et al. 2018. "Pandemics: Risks, Impacts, and Mitigation." In Disease Control Priorities: Improving Health and Reducing Poverty, edited by Dean T. Jamison, Hellen Gelband, Susan Horton, et al., 315–345. Washington, D.C.: The World Bank.
Maxmen, Amy, and Jeff Tollefson. 2020. "Two Decades of Pandemic War Games Failed to Account for Donald Trump." Nature 584(7819): 26–29.
Mearsheimer, John. 2001. The Tragedy of Great Power Politics. New York: W. W. Norton.
Meyer, Lukas H. 2003. "Past and Future: The Case for a Threshold Conception of Harm." In Rights, Culture, and the Law: Themes from the Legal and Political Philosophy of Joseph Raz, edited by Lukas H. Meyer, Stanley L. Paulson, and Thomas W. Pogge, 143–159. Oxford: Oxford University Press.
Meyer, Lukas H. 2020. "Intergenerational Justice." In The Stanford Encyclopedia of Philosophy (Summer 2020 Edition), edited by Edward N. Zalta. https://plato.stanford.edu/archives/sum2020/entries/justice-intergenerational.
Morgenthau, H. 1966. "Introduction." In A Working Peace System, edited by D. Mitrany, 7–11. Chicago: Quadrangle Books.
National Academy of Medicine. 2016. The Neglected Dimension of Global Security: A Framework to Counter Infectious Disease Crises. Washington, D.C.: The National Academies Press.
Nordgren, Anders. 2016. "Climate Change and National Self-Interest." Journal of Agricultural Environmental Ethics 29: 1043–1055.

Osterholm, Michael T. 2005. "Preparing for the Next Pandemic." *Foreign Affairs* 84(4): 24–37.
Parfit, D. 1976. "On Doing the Best for Our Children." In *Ethics and Population*, edited by M. D. Bayles, 100–115. Cambridge, MA: Schenkman Publishing Company Inc.
Parfit, D. 1984. *Reasons and Persons.* Oxford: Clarendon Press.
Paris Match. 2020. *Emmanuel Macron: "Nous sommes en guerre."* Paris, 16 March 2020. https://www.parismatch.com/Actu/Politique/Emmanuel-Macron-Nous-sommes-en-guerre-1678992.
Pike, Jamison, Tiffany Bogich, Sarah Elwood, et al. 2014. "Economic Optimization of a Global Strategy to Address the Pandemic Threat." *Proceedings of the National Academy of Sciences USA* 111(52): 18519–18523.
Posner, Eric A., and Cass R. Sunstein. 2008. "Climate Change Justice." *Georgetown Law Journal* 96(5): 1565–1612.
Poushter, Jacob, and Christine Huang. 2020. *Despite Pandemic, Many Europeans Still See Climate Change as Greatest Threat to Their Countries.* Washington: Pew Research Center, 9 September 2020. https://www.pewresearch.org/global/2020/09/09/despite-pandemic-many-europeans-still-see-climate-change-as-greatest-threat-to-their-countries/.
Ramsey, Frank. 1928. 'A Mathematical Theory of Savings." *The Economic Journal* 38(152): 543–559.
Rawls, John. 1971. *A Theory of Justice.* Cambridge, MA: Harvard University Press.
Raz, Joseph. 1994. "Rights and Individual Well-Being." In *Ethics in the Public Domain: Essays in the Morality of Law and Politics*, edited by Joseph Raz, 44–59. Oxford: Clarendon Press.
Robine, Jean-Marie, et al. 2008. "Death Toll Exceeded 70,000 in Europe During the Summer of 2003." *C. R. Biologies* 331: 171–178.
Samset, B. H., J. S. Fuglestvedt, and M. T. Lund. 2020. "Delayed Emergence of a Global Temperature Response After Emission Mitigation." *Nature Communications* 11: 3261.
Schreus, Miranda, and Elim Papadakis 2020. "Introduction." In *Historical Dictionary of the Green Movement*, edited by Miranda Schreurs and Elim Papadakis, 1–33. Lanham: Rowman & Littlefield.
Schwartz, Thomas. 1978. "Obligations to Posterity." In *Obligations to Future Generations,* edited by R. I. Siroka and Brian Barry, 3–13. Philadelphia: Temple University Press.
Shiffrin, Sheana. 1999, "Wrongful Life, Procreative Responsibility, and the Significance of Harm." *Legal Theory* 5(2): 117–148.
Shue, H. 1980. *Basic Rights: Subsistence, Affluence, and U.S. Foreign Policy.* Princeton: Princeton University Press.
Sidgwick, Henry. 1981 [1907]. *The Methods of Ethics* (7th ed.). Indianapolis: Hackett Publishing Company.
Steininger, Karl W., Lukas H. Meyer, Stefan Schleicher, Keywan Riahi, Keith Williges, and Florian Maczek. 2020. "Effort Sharing Among EU Member States: Green Deal Emission Reduction Targets for 2030." *Wegener Center Research*

Briefs 2-2020. Wegener Center Verlag, University of Graz, Austria, September 2020.

Stephens, Tim. 2016. "Disasters, International Environmental Law and the Anthropocene." In *Research Handbook on Disasters and International Law*, edited by Susan C. Breau and Katja L. H. Samuel, 153–176. Cheltenham: Edward Elgar Publishing.

Symons, Jonathan. 2018. "Realist Climate Ethics: Promoting Climate Ambition Within the Classical Realist Tradition." *Review of International Studies* 45(1): 141–160.

The White House. 2020. *Remarks by President Trump, Vice President Pence, and Members of the Coronavirus Task Force in Press Briefing.* Washington, 22 March 2020. https://www.whitehouse.gov/briefings-statements/remarks-president-trump-vice-president-pence-members-coronavirus-task-force-press-briefing-8/?utm_source=link.

Tollefson, Jeff. 2020. "Why Deforestation and Extinctions Make Pandemics More Likely." *Nature* 584(7820): 175–176.

Tromans, Stephen. 2010. *Nuclear Law: The Law Applying to Nuclear Installations and Radioactive Substances in its Historic Context.* London: Bloomsbury Publishing.

UNEP (United Nations Environment Programme). 2020. *Preventing the Next Pandemic: Zoonotic Diseases and How to Break the Chain of Transmission.* Nairobi: UNEP (The United Nations Environment Programme).

Waltz, Kenneth. 1979. *Theory of International Politics.* Berkeley: University of California.

Watts, Nick. 2019. "The 2019 Report of The Lancet Countdown on Health and Climate Change: Ensuring That the Health of a Child Born Today is Not Defined by a Changing Climate." *Lancet* 394: 1836–1878.

Weiss, Thomas G. 2013. *Global Governance: Why? What? Whither?* Cambridge: Polity Press.

Wendt, Alexander. 1992. "Anarchy is What States Make of It: The Social Construction of Power Politics." *International Organization* 46(2): 391–425.

WHO (World Health Organization). 2018. *Annual Review of Diseases Prioritized Under the Research and Development Blueprint Informal Consultation. Meeting Report.* Geneva: WHO. https://www.who.int/emergencies/diseases/2018prioritization-report.pdf?ua=1.

WMO (World Meteorological Organization). 2020. *Statement on the State of the Global Climate in 2019.* Geneva, No. 1248 (ISBN978-92-62-11248-5).

Working Group III to the Fifth Assessment Report of the Intergovernmental Panel on Climate Change. 2014. *Climate Change 2014: Mitigation of Climate Change.* Cambridge, MA: Cambridge University Press.

Worldometer [website]. https://www.worldometers.info/coronavirus/.

Wright, Robin. 2020. "The Coronavirus Pandemic is Now a Threat to National Security." *The New Yorker*, 8 October 2020. https://www.newyorker.com/news/our-columnists/america-the-infected-and-vulnerable.

Index

Ability to Pay Principle (APP), 4, 120, 122, 125
Abizadeh, Arash, 43, 44
action guiding, 16, 17, 25, 27, 131, 132, 134, 137–42, 146–48
activists, 39, 53, 55
adaptation gap, 41
agriculture, 69, 75
Aldred, Johnathan, 134
ambition, 3, 17, 18, 28, 61, 65, 84
ambitious theorizing, 94, 98, 102–6
ameliorative action, 178, 180, 182, 183
Andreou, Chrisoula, 205
anger (as a motivator), 176, 183
AP-NORC Center for Public Affairs Research, 194
APP. *See* Ability to Pay Principle
Assessment Report of the United Nations Intergovernmental Panel on Climate Change, 227
atomistic model, 72
Australia, 22

Baer, Paul, 16
Bailey, Alison, 162
basic rights, 223
Bayesian logic, 143, 144
Beardsworth, Richard, 233

BECCS. *See* bioenergy with carbon capture and storage
behavioral economics, 27
beneficiary pays principle, 4
Bergman, Anna-Karin, 99
Biden, Joe, 7
Binmore, Ken, 141
biodiversity, 2, 163
bioenergy with carbon capture and storage (BECCS), 162, 163
biomass, 70
Bolsonaro, Jair, 231
border control, 43, 220, 229
border regime, 43, 44
Brandstedt, Eric, 99
Brazil, 22
Brazil Proposal, 4
Brennan, Geoffrey, 98, 99
Brexit, 37, 40
Britain, 37
Broberg, Morten, 230
Broome, John, 8
building heating sector, 118

Canada, 22
Caney, Simon, 7, 8, 24, 117, 134
capitalism, 37
capitalist resource exploitation, 161
capturing carbon, 22

239

carbon budget, 21, 66–68, 70, 74
carbon capitalists, 139
carbon dioxide (CO_2), 18, 20–22, 65, 214, 219
carbon footprint, 117
carbon-intensive activities, 121
carbon pricing, 116, 118–23, 125, 126
carbon sinks, 215
carbon tax, 96, 117, 194
Carens, Joseph, 17
CC (climate change) mitigation, 194, 195, 198, 201, 202, 205, 207, 208. *See also* climate change
CDU. *See* German Christian Democratic Union
central government, 222, 223, 233. *See also* world government
CG. *See* climate goals
Chernobyl (accident), 229
Cherp, Aleh, 213
China, 21, 22, 230
Christian Social Union (CSU), 118, 120
circumstances of justice, 6
citizens, 10, 18, 104, 117, 123, 126, 137, 173, 178, 180, 182, 184, 194, 218, 220, 225, 228, 229
citizenship, 43
civil rights, 27, 28
civil society, 137, 213, 214, 230
clean energy, 69, 104
climate action, 1–4, 54, 55, 62–64, 79, 177, 178, 181, 185; constraints of, 9, 11, 174; feasibility, 8, 116, 180; and justice, 10, 122, 126; and self-interest, 72–75, 77, 186
Climate Behemoth, 36, 37
climate change, 2, 4, 6, 8, 11, 22, 24, 36, 49, 54, 61, 67, 71, 74, 75, 85, 100, 106, 124, 141, 146; anthropogenic, 155, 163–66; dangerous, 68, 70, 76, 78, 94, 98, 213, 214, 216, 220, 224, 229, 231, 233; feasibility, 10, 23, 101, 102, 117, 140, 174, 180, 181, 186; global, 1, 3, 5, 7, 163; impact of, 135, 136, 147, 148, 173, 183–85; justice/injustice, 9, 15, 16, 20, 28, 161, 165; mitigation, 119, 120, 125, 133, 155; and morality, 175–79, 182; policy, 25, 27, 43, 81
climate-change risk, 181
climate crisis, 7, 16, 28, 37, 71, 94, 137, 146, 173, 215, 226, 233; and feasibility, 10, 147, 186; and morality, 11, 135, 174, 181, 182, 185
climate denialism, 139
climate economics, 156
climate ethicists, 47, 48, 54, 82, 124–26, 175
climate ethics, 118, 123–26, 156
climate extortion, 9
climate goals (CG), 65, 70, 213, 215–17, 219, 222, 224, 226, 228, 230, 233
climate harm, 3, 5, 20, 22, 66, 165, 174–78, 183, 184, 227
Climate Leviathan, 36, 37
Climate Mao, 36, 37
climate models, 10, 143, 147, 148
climate policy, 5, 10, 19, 28, 34, 38, 41, 45, 47, 52, 54, 103
climate policy risk, 117
climate politics, 11, 131
climate-related risks, 3
climate rights, 99
climate scientists, 184
climate transition, 155, 156
climate treaty, 8, 9, 15, 19, 99, 101, 139, 156
Climate X, 36–38, 41, 45
clinical approach, 50–53, 55, 56n14, 57n22
CO_2. *See* carbon dioxide
coalitions, 11, 131, 137, 146, 181
cognitive biases, 39
cognitive capacity, 39
cognitive constraints, 39, 55
Cohen, Gerald, 23, 24, 55
cold-war militarism, 140
collaborative action, 179
collective responsibility, 178

colonialism, 28, 140
commercial planning, 148
common but differentiated responsibilities and respective capabilities, 4
communication of uncertainty, 184
compassion fade, 183
compatriots take priority principle, 225. *See also* my country first approach
compensation, 18, 134, 139
compliance constraints, 98, 99
Conference of the Parties (COP), 2
conservatives, 178, 184, 185
Convention on International Civic Aviation, 127n3
coronavirus, 213, 214, 221, 232. *See also* COVID-19
corporations, 96, 97, 103, 137
COVID-19, 7, 47, 48, 137, 193
criminal law, 25
critical civility, 168
critical race theory, 161
CSU. *See* Christian Social Union

Debating Climate Ethics (Weisbach), 62
decarbonization, 69, 71
deciding agents, 137
decision-making, 93, 95, 173, 179, 180, 182, 184, 186, 195, 215, 231
decision theory, 133. *See also* rational choice theory
deficit model, 164
deforestation, 22, 37, 218
Democratic Party (US), 7
Democratic Theory and Border Coercion, 43
descriptive norms, 175
desirability, 49, 132, 133, 135, 137
developing countries, 2, 20, 68, 139
diminishing marginal utility, 196
discounting, 196
discount rate, 20, 195–97
discursive kinds, 82
distributive justice, 4, 8, 15, 27, 155
dominated agents, 137, 146

Dunham, Jeremy, 162

economic models, 10
economic realists, 61, 73
economic wealth, 79
eco-socialism, 147
Edenhofer, Ottmar, 123, 124
efficacy, 138, 141, 177, 178, 180, 181, 186
electoral support, 117
electric vehicle, 22
emissions, 24, 28, 54, 67, 74, 75, 118, 136, 144, 145, 160, 202, 219; greenhouse gas, 1, 3, 8, 22, 37, 52, 115, 116, 133, 155, 175, 214, 221; individual, 199, 203, 205; policies, 101, 117, 122, 135; reduction, 4, 6, 10, 16, 18, 20, 21, 23, 61, 69, 95–97, 104, 120; zero, 62, 63, 65, 66, 78, 84, 193
emissions budget, 67. *See also* carbon budget
emissions egalitarianism (EE), 124
emissions rights, 20
emissions scenarios, 135
emotional down-regulation, 183
Energy Policy Institute, 194
energy transition, 62, 66, 69, 70, 84, 207
engagement, 49, 54, 147, 159, 173, 174, 178, 182, 185
enlightened self-interest, 232
environmental privilege, 163
environmental psychology, 121
epistemic, 134, 141, 142
equal per capita, 4, 121, 124
equity, 4, 149
Essays in Persuasion (Keynes), 233
Estlund, David, 17, 24–26, 34, 42
ethical concepts, 85
ethics-based approach, 63, 69. *See also* justice-based approach
EU (European Union), 6, 40, 116–18
Europe, 162, 214
European Union Emissions Trading System (EU ETS), 118
expected utility maximization, 133

ex post, 142
Extinction Rebellion, 67

fact-insensitive, 17, 23, 24
fallacy of approximation, 34, 42, 44, 54
familial inclusion model, 75
feasibility assessments, 38, 40, 45, 49, 56, 57, 131, 134–37, 140–42, 147, 148, 149n7, 158, 161, 179
feasibility claims, 16, 132, 159–62, 165–68
feasibility constraints, 5, 7, 11, 17, 26, 46, 53, 83, 118, 119, 158, 161; on climate action, 64, 84, 116, 174; on climate justice, 6, 9, 18, 19, 23, 27, 115, 126; and ideal theory, 123–25
feasible set, 50, 93, 94, 104, 105, 156
feminist philosophy, 83
Fifth Assessment Report, 4
first order responsibilities, 8
foreign aid, 19, 28
forest countries, 37
fossil fuels, 2, 22, 65, 69–71, 77, 84, 96, 104, 116, 118, 126, 137
France, 117
French government, 118
Fricker, Miranda, 55
Fukushima (accident), 229
future generations, 9, 40, 46, 95, 97, 184, 226–28, 231, 232

Galston, William, 25
Gardiner, Stephen, 5, 9, 16, 19, 24, 40, 41, 177, 226
Gaus, Gerald, 35, 51
general theory of second best, 105
geopolitical constraints, 215–17
German Christian Democratic Union (CDU), 118, 120
German government, 118
Germany, 81, 118, 120, 125, 166
Geuss, Raymond, 55
Gheaus, Anca, 27
GHG. *See* greenhouse gases
Gifford, Robert, 181

Gilabert, Pablo, 103, 133, 141, 228
global climate regime, 4
global constitutional convention, 40
global economy, 37
global inclusion model, 76
global justice, 115, 140
global mean surface temperature (GMST), 135, 136, 143–45
Global North, 155, 161–63
global poor, 96, 97, 137, 139, 140, 146
Global Preparedness Monitoring Board, 230
Global South, 163
global warming, 1–3, 183
GMST. *See* global mean surface temperature
Great Lakes region, 147
green economy, 155
green future, 155
Greenhouse Development Rights, 16
greenhouse gases (GHG), 1–3, 5, 8, 21, 66, 95, 115, 116, 118, 120, 135, 136, 144, 161
green new deal, 126, 135, 136, 155
Green Party, 120, 121

Hacking, Ian, 82
hard constraints, 38, 84, 214–16, 224
Heath, Yuko, 181
heatwave, 214
hermeneutically deprived, 164
high-emitting countries, 96
Hiroshima (bombing), 229
humanitarian aid, 221
Hurricane Florence, 183
Hurricane Harvey, 183

ideal benchmarks approach, 35, 36, 42–45, 54
ideal of justice, 132
ideal society, 33, 36, 38
ideal targets approach, 35, 36, 38, 40, 42, 45, 51, 52
ideal theorizing, 123, 132, 156
ideal theory, 6, 98, 115, 116, 123–26, 132

identifiable victim effect, 183
immediacy multiplier, 196, 197, 203
immigration, 43, 44
immoral behavior, 179
incentive compatibility, 100, 108n17
indefinite chain inclusion model, 76
India, 21, 22
industrialized nations, 4
inequality, 27
injustice, 6, 10, 16, 48, 52, 53, 102, 121–26, 155, 157, 159–61, 163–68
institutional reform, 55
integrated assessment models, 157
intergenerational cooperation, 4, 217, 219, 220
intergenerational ethics, 77
intergenerational justice, 115. *See also* global justice
Intergovernmental Panel on Climate Change (IPCC), 3, 16, 22, 30, 68, 70, 94, 135, 144, 214, 227
International Atomic Energy Agency (IAEA), 229
international climate change agreement, 124
international cooperation, 40, 221, 229
international justice, 6
international law, 83, 228
international negotiations, 1
International Paretianism (IP), 8, 19, 99–101
interpretive climate ethics, 34, 47
intertemporal preference reversal, 197, 198, 200–6
intransitive preferences, 200, 204, 206
IP. *See* International Paretianism
IPCC. *See* Intergovernmental Panel on Climate Change
IPCC Assessment Report (2014), 47
iron law of wealth, 79

Jamieson, Dale, 5
Jewell, Jessica, 213
Johnson, Boris, 231
justice-based approach, 64, 84

justice-improvement, 6, 44
just transition, 155, 168

Keynes, John Maynard (*Essays in Persuasion*), 233
Kowarsch, Martin, 123, 124
Kyoto Protocol, 2, 4, 82

Lawford-Smith, Holly, 102, 133, 141, 162, 228
The Law of Peoples (Rawls), 124
least developed countries (LDCs), 139, 140
legitimate expectations, 161
lexical priority, 167
liberal institutionalism, 221, 222, 224
liberals, 178, 184
low-carbon economy, 134
low carbon society, 125

Macron, Emmanuel, 213
Madison, James, 18
male privilege, 161
Mann, Geoff, 36–38
maximum individual utility risks, 200
max tax, 17, 24
McBride, Cillian, 164
McIntosh, Peggy, 162
McTernan, Emily, 158
Mearsheimer, John, 222, 223
Merkel, Angela, 119
method of Lagrange multipliers, 26
migration governance regime, 44
militarized adaptation, 149
military, 76, 131, 148
Mills, Charles, 53
mindset of a fierce optimism, 233
Mirrlees, James, 17
mitigation, 16, 75, 95, 97, 98, 102, 104, 131, 144, 156, 166, 199, 214, 226, 230; adaptation goals, 216, 218, 219, 231; climate change, 7, 28, 66, 125, 161, 162, 194, 195, 198, 201, 202, 205, 207, 208; policies, 103, 117, 119, 120, 134, 135

moral action, 176, 177
moral decision-making, 104, 185
moral disengagement, 178
moral expansionism, 185
moral ideals, 182, 184
moral intuitions, 174, 176
moralizing language, 185
moral judgment, 177, 179, 180, 182, 186
morally relevant action, 174
morally relevant behavior, 11
morally relevant domains, 176
morally relevant features, 36
moral motivation, 159, 183, 194
moral norms, 175, 176
moral outrage, 183
moral reactions, 174, 177, 183
moral urgency, 7, 202
moral values, 184, 185
moral violations, 183
Morgenthau, Hans, 223, 224
Morrow, David (*Values in Climate Policy*), 47
motivational gap, 194, 201
motivations, 16, 18, 19, 22, 24, 25, 27, 97, 103, 118, 121, 158, 166, 182, 201; lack of, 17, 117, 122, 178; moral, 159, 180, 183, 194, 208
multilateral climate agreement, 2, 43
multilateral foreign aid treaty, 28
mutual deterrence, 224
my country first approach, 225, 229, 231

Nagasaki (bombing), 229
naive procrastinator, 201, 202
National Academy of Medicine, 229
national inclusion model, 76
nationalism, 225, 226
nationally determined contributions (NDC), 2, 3
national security, 149
national wealth, 80
natural kinds, 82

NDC. *See* nationally determined contributions
neo-liberalization, 140
net zero emissions, 67
neutral preferences, 195
NGO (non-government officials), 53, 55, 118
nomological constraints, 88n25
non-compliance, 6
non-economic disutility, 207
non-ideal circumstances, 124
non-ideal theory, 6, 47, 51, 54, 116, 123–26
non-identity-problem, 227
normative claims, 23, 24, 100
normative goals, 17
normative literature, 4, 158
normative policy discourses, 156
normative reasons, 196
normative theory, 2, 5, 7, 8, 10, 11, 93, 138, 157, 233
norm violations, 175, 177
Nozick, Robert, 26
nuclear attack, 224
nuclear energy, 229
nuclear war, 224
nuclear weapons, 224
Nussbaum, Martha, 27

obstructionism, 83
100-year horizon, 66, 67, 69
OPEC (Organization of the Petroleum Exporting Countries), 22
opposing agents, 137, 138, 139
optimistic hypothesis, 232, 233
Osterholm, Michael, 218
ought implies can principle, 98
outcomes, 17, 18, 27, 28, 55, 74, 93, 98, 102, 103, 105, 132, 134, 138, 141, 147, 148, 175, 180

pandemic, 40
pandemic goals, 216, 217, 220, 224, 228
Parfit, Derek, 226

Paris Accord, 139. *See also* Paris Agreement
Paris Agreement (2015), 1–4, 7, 21, 29n8, 40, 115, 118
past emissions, 16, 18, 24, 123
path-dependency, 88n27
personal responsibility, 178, 199
Pettit, Philip, 98, 99
Pew Research Center, 194
PG. *See* pandemics goals
philosophers, 5, 15, 25, 33, 48, 82, 93, 195, 226; and climate policy, 46, 47, 53–55; political, 1, 98, 149; and utopian approach, 34, 38, 41, 45
Plato, 24
pluralism, 138, 146
polarization, 178, 183
policymakers, 4, 36–39, 47, 50, 124, 162; and climate justice, 10, 99, 100, 186; feasibility constraints on, 34, 48, 53; ideal approach on, 40–42; and philosophers, 33, 45, 46, 52, 54, 55
polio, 221
political legitimacy, 43
political philosophy, 6, 15, 16, 124, 132, 157
political theory, 156, 157
political traction, 99
politics of the armed lifeboat, 149
Polluter Pays Principle (PPP), 4, 119, 120, 122, 125
positive argument, 61–64, 67, 72, 74, 78, 84, 85
Posner, Eric, 99, 101, 124, 156
possibilistic practical reasoning, 146, 148, 149
possibility theory, 143
poverty, 2, 96, 106
power, 179, 222, 223
power imbalances, 159
power inequities, 146, 226
power plants, 65
power structures, 11, 157, 160
PPP. *See* Polluter Pays Principle

pragmatic constraint, 101
pre-commitment, 202–4
preference reversal, 197, 198, 202
principles of distributive justice, 1
privilege, 11, 157, 161–68, 173
probabilistic feasibility assessment, 131, 133–37, 140–42, 147, 148
probability assessment, 10, 132
probability judgments, 144
probability statements, 144, 145
probability theory, 143
procrastination, 85
procrastination containment mechanism, 206
procrastination loop, 197–200, 202–5
procrastinators, 201
property rights, 162
proximity constraints, 215, 217, 220, 225, 226, 228, 229, 231, 233
psychological barriers, 174
psychological mechanisms, 11, 174, 178, 183, 201
psychology, 5, 11, 185, 215
public health, 7
public opinion, 39, 82, 183
public reason, 169n4

rainforest, 37
raising the costs, 205
Ramsey, Frank, 227
rational choice theory, 133–35, 137, 138, 139, 140–42, 148
Rawls, John (*The Law of Peoples; A Theory of Justice*), 33, 38, 123–25, 132, 133, 227
RCP. *See* representative concentration pathways
realism, 17, 62, 64, 84, 221–24, 229, 231–33
realistic possibilities, 143, 144, 147, 148
realistic utopia, 6
reciprocity, 184
reflexivity, 146, 147
reframing behavior, 185
remediable injustice, 48

renewable energy, 69, 96, 104
representative concentration pathways (RCPs), 135, 144, 145
rights-based theories, 26
right-wing nationalist, 37, 229
Rio Earth Summit, 2
robust strategies, 146
Roser, Dominic, 141–45, 156, 159
Russia, 21. *See also* Soviet Union

SARS-CoV-2, 213
Savage, Leonard, 141
Sayegh, Alexandre Gajevic, 99, 134
scenario analysis, 146, 148
Schwartz, Thomas, 226
second-order procrastination, 205
second order responsibilities, 8
self-interest approach, 61, 62, 68, 69, 72
self-interest collective, 76
Sen, Amartya, 26, 50, 51
sexual harassment, 55
Shue, Henry, 225
Sidgwick, Henry, 227
Simmons, A. John, 26, 50, 51
simple self-interest argument, 64, 69, 71, 85
Singer, Peter, 16, 25
skepticism, 93, 94, 138, 227
smallpox, 221
Social Democratic Party of Germany (SPD), 118, 120
social discount rate, 196
social justice, 116, 118, 122
social order, 39–42, 46, 50, 56
social recognition, 164
socio-economic, 155, 163
socio-political discourse, 167, 168
soft constraints, 115, 214–17, 219, 220, 224, 228
solar panels, 175, 203, 206
solar (power), 22
sophisticated procrastinator, 201, 202, 204
Soviet Union, 136
space-related proximity constraints, 217

SPD. *See* Social Democratic Party of Germany
Special Report on Global Warming of 1.5°C, 3
status quo, 52, 75, 83, 84, 160–63, 165–68, 179, 182, 200
status quo bias, 53, 207
sticky institutional structures, 6
strategic climate planning, 131, 147
strong-IGC (intergenerational cooperation XE "intergenerational cooperation"), 219, 220, 225, 226, 231
strong-INC (international cooperation), 218–23, 225, 226, 229–32
structural constraints, 215, 217, 218, 220–22, 224, 229, 231, 232
structural injustice, 159, 161, 166
sub-Saharan Africa, 162, 163
subsidies, 207
sustainable development, 155
sustainable transition, 155

Tank, Lukas, 119, 120
taxes, 17, 18, 26–28, 48, 96, 104, 116–18, 120–22, 125, 160, 194, 207
technological feasibility, 157
temporal discounting, 184
temporal proximity constraints, 228
A Theory of Justice (Rawls), 124, 227
theory of rights, 227
threat multipliers, 232
three generations, 71, 72, 75, 76, 84
time-related proximity constraint, 217, 218
trade agreement, 37
trade union movement, 155
transitional pathways, 138
transitivity of preferences, 198
transportation, 118, 205, 224
Tribal communities, 147
true preferences, 11, 195, 197, 198, 202–4, 208
Trump, Donald, 37, 40, 213, 231, 232
Tsebelis, George, 138

unacceptable harms, 86n13
uncertainty (concept), 36, 38, 65, 68, 141, 143, 144, 145, 148, 177, 183–86; of climate change, 81, 134, 135, 136, 146, 178, 222; and feasibility, 40, 41, 45, 133
UNFCCC. *See* United Nations Framework Convention on Climate Change
United Nations, 37
United Nations Framework Convention on Climate Change (UNFCCC), 2, 4
United States, 2, 7, 18, 21, 37, 40, 52, 77, 82, 96, 101, 126, 135, 139, 163, 178, 179, 229, 231
United States Department of Defense, 232
United States Intelligence Community, 230
unrealistic possibilities, 143, 144
unwarranted privilege, 159
U.S. Congress, 138
U.S. Constitution, 18
U.S. Joint Chiefs of Staff, 232
U.S. Senate, 7, 52, 230
utilitarianism, 227
utility multiplier, 201, 203
utopian approach, 33–36, 42, 45, 46, 50–52, 54
Utset, Manuel, 196, 201

vaccine nationalism, 220
vaccines, 218, 220, 226
Valentini, Laura, 26
Values in Climate Policy (Morrow), 47

vulnerability, 137, 139, 140, 147, 163, 166, 232

Wainwright, Joel, 36–38
Waltz, Kenneth, 221
wealth, 79
wealth redistribution, 99, 121, 193
wealthy citizens, 104
wealthy countries, 20, 24, 96
wealthy individuals, 96, 103, 179
wealthy nations, 124
Weisbach, David (*Debating Climate Ethics*), 9, 65, 79–81, 83, 85, 124, 156; International Paretianism, 8, 99, 101; positive argument, 61–64; self-interest, 10, 69–75, 77, 83; zero emissions, 66–68, 78
Weitzman, Martin, 21
white privilege, 161, 163
WHO. *See* World Health Organization
Whyte, Kyle Powys, 147
Wiens, David, 38, 47–50, 52
Window, Overton, 27
Wind (power), 22
world government, 40, 221, 222
World Health Organization (WHO), 40, 218, 221, 229
World Meteorological Organization, 214

Yale Program on Climate Change Communication, 194
yellow-jacket movement, 10, 117
Young, Iris Marion, 161

zero emissions, 22, 62, 65–68, 76, 78, 84, 193

About the Contributors

Marcelo de Araujo is professor of ethics at the State University of Rio de Janeiro, and professor of philosophy of law at the Federal University of Rio de Janeiro. He received his doctorate in philosophy from the University of Konstanz, Germany, in 2002. He was a Humboldt Research Fellow at the University of Konstanz (2007–2008, 2014, 2018) and visiting scholar at the Uehiro Centre for Practical Ethics, Oxford (2013). He was also a visiting professor at the Master EELP (Ethics—Economics, Law and Politics) at the University of Bochum, Germany (2019–2020). His current philosophical research, funded by the CNPq (National Council for Scientific and Technological Development, Brazil) and by FAPERJ (Research Support Foundation of the State of Rio de Janeiro, Brazil), focuses on the ethical and legal implications that arise in the quest for human enhancement, and on the contemporary debate on climate ethics.

Fausto Corvino is a postdoctoral research fellow in theoretical philosophy in the Department of Philosophy and Educational Sciences at the University of Turin. He is also member of Labont (Center for Ontology) at the University of Turin and Affiliate Researcher in the DIRPOLIS Institute (Law, Politics and Development) at Sant'Anna School of Advanced Studies in Pisa. Prior to this, he was a postdoctoral research fellow in moral philosophy and member of the Research Group on Public Ethics (2018-2020) at Sant'Anna School of Advanced Studies, where he was awarded a PhD (2017) in politics, human rights, and sustainability (with a research curriculum in political theory). His main research interests lie in theories of distributive justice (including global and intergenerational justice), philosophy of economics, and applied ethics (including the ethics of climate change).

Stephen M. Gardiner is professor of philosophy and Ben Rabinowitz Endowed Professor of the Human Dimensions of the Environment at the University of Washington, Seattle, where he is also director of the Program on Ethics. His research focuses on global environmental problems, intergenerational ethics, and virtue ethics. He is the author of *A Perfect Moral Storm: The Ethical Tragedy of Climate Change* (Oxford, 2011), coauthor of *Debating Climate Ethics* (Oxford, 2016), editor of *Virtue Ethics, Old and New* (Cornell, 2005) and the *Oxford Handbook of Intergenerational Ethics* (Oxford, under contract), and coeditor of the *Oxford Handbook of Environmental Ethics* (Oxford, 2016), *Climate Ethics: Essential Readings* (Oxford, 2010), and *The Ethics of "Geoengineering" the Global Climate: Justice, Legitimacy and Governance* (Routledge, 2020). His articles have appeared in journals such as *Ethics, Ethics and International Affairs, Journal of Political Philosophy, Oxford Studies in Ancient Philosophy,* and *Philosophy and Public Affairs*.

Jared Houston earned a bachelor of science in mechanical engineering from the University of Calgary and then worked as a petroleum engineer for four years before pursuing studies in philosophy. He graduated with a PhD in philosophy from Queens University (Canada) in 2019. His research focuses on political theory and climate change, including work appearing in the *Radical Philosophy Review*.

Corey Katz is an assistant professor of philosophy at Georgian Court University in Lakewood, NJ. He was a postdoctoral researcher in the ethics of sustainable development at the Ohio State University from 2016 to 2018 and received his PhD from Saint Louis University. He has published on how we should understand our moral responsibilities to future generations, and on the ethics of greening healthcare, with papers appearing in *Ethical Theory and Moral Practice, Ethics, Policy & Environment,* and the *Journal of Medicine and Philosophy*. His research interests also include Kantian contractualism and the question of individual responsibility for overdetermined harms like climate change.

Sarah Kenehan is an associate professor of philosophy in the Department of Philosophy and Religious Studies at Marywood University in Scranton, PA. She earned her PhD in 2010 from Graz University in Austria, with specializations in liberal political philosophy (esp. Rawls), international justice, and climate justice. She currently teaches and writes in various areas of ethics and social and political philosophy, including climate justice, Rawls, and animal ethics. Most recently, she coedited (with Erinn Gilson) and contributed to the volume *Food, Environment, and Climate Change: Justice at the*

Intersections (RLI, 2019). Other recent work has appeared in *Climate Justice and Historical Emissions* (eds. L. Meyer and P. Sanklecha, Cambridge, 2017) and *The Ethics of Animal Experimentation: Working Towards a Paradigm Change* (eds. K. Herrmann and K. Jayne, Brill, 2018).

Justin Lawson is a doctoral candidate in philosophy at the University of Washington. He studies topics at the intersection of philosophy of science, political philosophy, and environmental ethics.

Ezra Markowitz is an associate professor of environmental decision-making in the Department of Environmental Conservation at the University of Massachusetts Amherst. His research and teaching focus on the intersection of decision-making, persuasive communication, public engagement with science, and environmental sustainability. He is particularly interested and expert in the practical application of behavioral science to improve individuals' and communities' environmental decision-making; he also has deep expertise in the field of climate change communication and public engagement. He is the author of over five dozen peer-reviewed research papers, book chapters, and reports, including the 2015 *Connecting on Climate* guide to climate-change communication. At UMass Amherst, Markowitz teaches courses on environmental decision-making, conservation social science, and public engagement and communication for scientists. He holds a PhD in environmental sciences, studies and policy, an MS in psychology from the University of Oregon, and a BA in psychology from Vassar College.

Joshua D. McBee is an associate at Climate Advisers in Washington, DC, where he focuses on advancing natural climate solutions. Prior to joining Climate Advisers, he completed a PhD in philosophy at Johns Hopkins University. At Hopkins, he taught climate ethics and wrote a dissertation that defends a deflationary metaethics and advances a novel, Wittgensteinian response to evolutionary debunking arguments.

Kirsten Meyer studied philosophy and biology at the universities of Münster, Bielefeld, and St. Andrews and obtained a PhD in philosophy. She was an assistant professor of philosophy at the Universities of Regensburg and Göttingen, and a junior professor in Berlin. She has been a full professor of practical philosophy at Humboldt University since 2011. Her research areas include normative ethics, applied ethics, and political philosophy. The main topics of her research are the ethics of future generations, the ethics of nature, climate ethics, and the philosophy of education. Among her publications are the books *Der Wert der Natur (The Value of Nature)* 2003, *Bildung* 2011, and *Was schulden wir künftigen Generationen? (What do we owe to future generations?)* 2018.

Lukas H. Meyer is professor of philosophy at the University of Graz, Austria, and speaker of the Working Unit Moral and Political Philosophy. He received his doctorate from Oxford University and completed his habilitation at the University of Bremen, was faculty fellow in ethics at Harvard University (2000–2001), Humboldt Research Fellow at Columbia University (2001–2002), and assistant professor of philosophy at the University of Berne (2005–2009). Since 2014, he has been the speaker of the interdisciplinary Doctoral Program Climate Change (funded by the Austrian Science Fund), and since 2019, the speaker of the inter-faculty Field of Excellence Climate Change Graz. He served as dean of the Faculty of Arts and Humanities at Graz from 2013–2017. As one of the first philosophers to work in the organization, Meyer served as a lead author for the Intergovernmental Panel on Climate Change (IPCC). His book publications include *Climate Change and Historical Emissions* (co-edited, CUP 2017); *Intergenerational Justice* (co-edited, OUP 2012); *Legitimacy, Justice and Public International Law* (edited, CUP 2009); and *Historische Gerechtigkeit (Historical Justice)* (de Gruyter 2005). He is a founding editor of the journal *Moral Philosophy and Politics* (de Gruyter). For more information see https://homepage.uni-graz.at/en/lukas.meyer/ and https://online.uni-graz.at/kfu_online/wbforschungsportal.cbshowportal?pPersonNr=69399

Andrew Monroe is an associate professor in the Department of Psychology at Appalachian State University. His research follows two main tracks. First, he studies how personal moral values and beliefs influence social action on culturally hot issues (e.g., support for the BLM movement & LGBTQ rights) and second, he studies how and when people communicate moral sentiments (blame and praise), and identifying the relational and social dynamics that shape people's willingness to express and update moral judgments. He has published over two dozen peer-reviewed research articles and currently teaches classes on moral psychology, social cognition, and social psychology at Appalachian State University. He holds a PhD in psychology from Brown University.

Fabian Schuppert is professor of political theory at the University of Potsdam, Germany. Before coming to Potsdam, Fabian held positions at Queen's University Belfast and the Centre for Ethics at the University of Zurich. Fabian's primary research interests lie in theories of social equality, republicanism, democratic socialism, risk ethics, and climate ethics. Currently, Fabian is working on a project on land rights in the context of climate change, with a special view to decolonizing existing climate justice discourses, as well as a project on systemic risks and their ethical governance.

David Weisbach is the Walter J. Blum Professor at the University of Chicago Law School. He writes and speaks regularly on climate change and on tax policy. He has degrees in mathematics from the University of Michigan and Wolfson College Cambridge, and a law degree from Harvard. Before joining academia, he worked in private legal practice and at the Office of Tax Policy in the US Department of the Treasury. He was previously a senior fellow at the University of Chicago Computation Institute and Argonne National Laboratories, and is currently an international research fellow at the Said School of Business, Oxford University. He was elected to the American Academy of Arts and Science in 2012 and the American Law Institute in 2008.

www.ingramcontent.com/pod-product-compliance
Lightning Source LLC
Chambersburg PA
CBHW021848300426
44115CB00005B/57